J. W. Slater

Sewage Treatment, Purification, and Utilization

J. W. Slater

Sewage Treatment, Purification, and Utilization

ISBN/EAN: 9783743390126

Manufactured in Europe, USA, Canada, Australia, Japa

Cover: Foto ©Lupo / pixelio.de

Manufactured and distributed by brebook publishing software (www.brebook.com)

J. W. Slater

Sewage Treatment, Purification, and Utilization

SEWAGE TREATMENT, PURIFICATION, AND UTILIZATION.

A
PRACTICAL MANUAL FOR THE USE OF CORPORATIONS,
LOCAL BOARDS, MEDICAL OFFICERS OF HEALTH,
INSPECTORS OF NUISANCES, CHEMISTS,
MANUFACTURERS, RIPARIAN OWNERS,
ENGINEERS AND RATEPAYERS.

BY

J. W. SLATER, F.E.S.

LATE EDITOR OF "JOURNAL OF SCIENCE." AUTHOR OF "MANUAL OF COLOURS AND DYE WARES."

LONDON:
WHITTAKER & CO., PATERNOSTER SQUARE.
GEORGE BELL & SONS, YORK STREET, COVENT GARDEN.
1888.

(The right of Translation is reserved.)

PREFACE.

THE importance of the Sewage question may, perhaps, be fairly gauged by the number of patents of which it has been the subject, and its still unsettled state may be concluded from their varied and contradictory character.

Freezing and heating, concentration and dilution, electrisation and magnetising, the addition of oxidisers and deoxidisers, of ferments and preventives of fermentation recommended, if not actually tried, show the want of any distinct and generally recognised principle.

I have, therefore, thought it my duty to lay before the public in plain language and in a concise form the result of my experience in treating sewage and waste waters. That experience dates back to the year 1868, and except in 1870 my attention has during the whole of that time been almost exclusively directed to this question. I have worked with sewage in every quantity, from a few ounces in a beaker or a hydrometer-jar to a daily flow of ten million gallons. I have had opportunities for examining almost every known process, irrigation, filtration, aeration as well as precipitation in its many variations. I have studied sewage in droughts and in storms, in hot weather and in cold, by day and by night, in residential towns and in industrial centres. My conclusion is that there is no one process universally applicable.

Unfortunately there is no subject, outside the range of party politics, on which so much envy, hatred, malice, and all uncharitableness prevail as on the treatment of sewage. But I ask people to judge by the evidence of their own senses. Do not read about this or that process, but go and look. I know instances where bitter enemies of chemical processes have been convinced of their error by just one unexpected and unprepared for visit of inspection.

A few supplementary remarks and explanations are needful :—

The list of patents is not by any means as complete as the author could have wished. Under the existing state of the Patent law nearly twelve months may elapse from the date of application to that of the acceptation of the complete specification, and more time elapses before such "complete" is printed and is to be found in its place in the Patent Office Library. In the meantime no one can tell whether a patent has been abandoned or not. The difficulty is increased by the circumstance that provisional specifications, if not proceeded with, are now not printed, as heretofore. This renders it difficult to obtain a correct list of the patents for any given year until about eighteen months have elapsed from its termination. When opposite the number of a patent there appear the words "not published," or a blank space, the reader will understand that the "complete" was not to be found at the dates when the author made his searches.

The specification M.L.G.G. Daudinart (A.D. 1886),

No. 4,203, was misunderstood. The zinc precipitate is not applied as a manure.

It should be noted that, if we may judge from the local papers, the process adopted at Hendon (p. 91) gives but a very limited degree of satisfaction to the ratepayers.

As regards "germs" or morbific ferments, it is now generally held that these tiny organisms when introduced into the system are not the direct causes of disease and death, but that they generate within the body they invade certain most intense poisons, which do the deadly work. Practically speaking this is not a matter of importance. If we can prevent the entrance of these "germs" into our system, or if we can destroy them after entering, they can have no opportunity to develop poisons.

Here comes up another point:—In addition to the disease-germs, "pathogenic organisms," as they are technically called, there exist in sewage and in other waters germs of a very different class. These latter it would seem, according to the researches of Dr. Dupré, play a very important part in the purification of waters. If, therefore, we remove all germs, whether by filtration, precipitation, disinfection, or any other process, we may in some cases do more harm than good. Can we, therefore, devise some discriminating means which shall destroy the evil germs and leave the useful ones untouched? If we can do this we may hope to render sewage potable. Here, then, is scope for the inventor.

The action of electric currents upon polluted waters remains to be studied. Laboratory experiments hold out hopes, but on the large scale they may prove too costly.

The shape, size, and general construction of tanks, (p. 111) likewise require further study. I may here record my conviction that, wherever practicable, intermittent working is preferable to continuous treatment. The sewage is better dealt with at a less outlay for power, chemicals, and labour.

There is surely room and need for that fair cooperation of experts which Dr. Dupré asks for, as quoted on p. 267.

Journal of Science Office,
Ludgate Hill, E.C.

CONTENTS.

INTRODUCTION.

CHAPTER I.

Nature and Composition of Sewage—Sewage of Residential Towns—of Manufacturing Towns—Day and Night Sewage—Sunday Sewage—Water Supply—Single or Double Sewerage . . . . Pages 1—6

CHAPTER II.

How, When and Where is Sewage Injurious—Definition of "Nuisance"—Microbia in Sewage—Sewage Detrimental to Fish—Organic Nitrogen in Sewage—Urea, Ammonia, Nitrogen Acids—Phosphoric Acid—Potash, Common Salt—Putrescent Vegetable Matter—Sulphur Compounds—"Hardness" of Sewage—its Temperature—Sewer Ventilators—Matters Hurtful to Fish—Hurtful to Plants . . . . . . 9—24

CHAPTER III.

Disposal and Treatment of Sewage—The Cesspool—The Dry Closet—Pneumatic Systems . . . . 24—30

CHAPTER IV.

Water-carriage in General—First use of Water-closet—Sewer Gas—Dangers of Escape into Houses—Proposals for dealing with Sewer Gas by P. Spence and W. A. Gibbs—The Dust Bin . . . . 30—37

CHAPTER V.

The Bazalgette System a Failure in London . Pages 37—43

CHAPTER VI.

Irrigation, its Principle and Conditions—Quality of Soil and Climate—Italian Rye-Grass convertible into Hay—Irrigation at Gennevilliers—Soil Water-proofed—Manurial Matter in Sewage only Partially Utilised—Irrigation does not remove "Germs"—Encourages Flies—Flies convey Disease-Germs—Experiments of Mr. Smee, Jun. . . . . 43—67

CHAPTER VII.

Modifications of Irrigation — Settling-pits — Introduction of Disinfectants — Irrigation as Supplement to Precipitation . . . . 67—71

CHAPTER VIII.

Filtration—Structure of Filter-beds—Filtering-materials—Spongy Iron—Professor H. Robinson's Process—Intermittent Downwards Filtration . . . . 71—82

CHAPTER IX.

Precipitation—"Clarify and Purify"—Occlusion or Absorption—Organic matters eliminated by Chemical Processes—Properties required for Precipitants—Lead Condemned—Iron—Alkaline Effluents — Lime condemned — Sulphates not Desirable — Gypsum condemned — Sulphates, Hypochlorites, Salts of Barium condemned — Useless Agents — Phosphate Processes—Salts of Aluminium—Alum not Suitable—Sodium Aluminate—Salts of Manganese—Salts of Zinc and Copper—Absorbents, Fatty Clays, Coke, Peat, Charcoals, Gelatinous Silica—Joint action of Precipitants and Absorbents—Inverse Irrigation—Intermittent or Continuous Working—Treatment of Sewage Sludge—Knostropp Works—Clean Tanks—Supplementary Filtration—General Arrangements of Works—Bogus Patents and Bogus Working . . . . . 83—118

CHAPTER X.

Deodorising— Chloride of Lime—Manganates—Aëration of Effluents and of Sewage — Gases patented for Treatment of Sewage . . . . Pages 118—124

CHAPTER XI.

Destruction—The Scott Process—Sewage Cement . 124—126

CHAPTER XII.

Promiscuous Methods — Distillation — Freezing — Electricity . . . . 127—133

CHAPTER XIII.

Self-Purification — Cellular Chemical Treatment — Rivers Self-purifying—The Vesle, the Calder and Hebble Navigation—The Passaic, the Oder, the Seine . . . . . 134—142

CHAPTER XIV.

Detection of Sewage Pollution in Rivers - Watercress in Polluted Waters — Experiments—Sewage Fungus—Professor Koch on Microbia in Water—Dr. Link on the Microscopic examination of Water—Pollution of Wells—*The* "Recommendations"— Their Shortcomings—Rational Scale . . . 143—163

CHAPTER XV.

Recognition of Purification in Sewage Tanks—Underground Pollution—Sampling . . . . . . 164—171

CHAPTER XVI.

Precipitation Mud, methods of Drying—Stanks—Filter Presses— Press Liquor—Drying Cylinders . . . . 178—187

CHAPTER XVII.

Sewage Manures—Errors regarding . . . . 172—178

CHAPTER XVIII.

Sewage Legislation—The Act of 1876—Its Defects—Proposed Measure of 1885 . . . . . Pages 186—190

CHAPTER XIX.

Sewage Patents . . . . 191—256

CHAPTER XX.

Discussion on Dr. Tidy's Paper, read before the Society of Arts May 5th, 1886 . . . . 257—267

SEWAGE TREATMENT, PURIFICATION, AND UTILISATION.

CHAPTER I.

NATURE AND COMPOSITION OF SEWAGE.

STRANGE to say, even professed sanitarians, municipal authorities, and the like—not to speak of educated and intelligent people in general—have often very vague notions concerning the nature of sewage. They know that it is nasty, and that it contains sundry matters unsightly and offensive. But they are little aware of its complexity, and of its liability to vary according to the locality, the season, the weather, and even the hour of the day. Were these points better understood many futile processes for sewage treatment would never have been proposed; or, if proposed, would never have been entertained.

We will take, first, the simplest case, that of a "residential" town. By this term sanitarians mean a town where few, if any, manufacturing operations are carried on, and where the liquid refuse is mainly of domestic origin. As examples, we may mention Oxford, Winchester, Bury St. Edmunds, Keswick, Leamington, and Aylesbury. The sewage of such towns is generally of a concentrated character, and the matters which it holds in solution are, for the most part, of an organic and readily putrescible nature. It contains the solid and liquid excretions of the inhabitants, the urine of horses and cattle discharged in the streets, the drainage of stables and piggeries, the blood (more or less) of cattle slaughtered in the town, and the washings of the slaughter-houses. Another important constituent is the water which has served for washing the persons of the inhabitants, their clothing, and their cooking utensils, etc. These " slops," as they are called, are very offensive ; they hold in suspension and solution soap, fatty acids, the juices of meat and vegetables, and the exudations from the human skin. Almost every one must have observed that if a bowl of suds from "washing-day" has been allowed to stand, it gives off in the course of a few hours a most unpleasant odour. There are also certain organic substances not offensive nor readily capable of putrefaction, but which, nevertheless, play a most important part as far as the treatment of sewage is concerned. Thus a very large quantity of waste paper finds its way into the sewers, and is there subdivided into particles, which quite escape the notice of an inexperienced observer. To these are added fine filaments derived from washing linen and cotton articles, and, to a less extent, from woollens. I say to a less extent, not because woollens when washed give off a smaller quantity of fibre, but because the total amount of woollens washed in a residential town is much smaller than that of cotton and linen fabrics. These filaments of textile materials and of paper are scarcely perceptible on a hasty examination. A glass of sewage held up to the light, if not rendered turbid by suspended clay, road-silt, etc., appears almost as limpid as ordinary river or pond water. But if we try to filter it through the filter-paper used by chemists, we find that the flow, though tolerably rapid at first, quickly slackens, and soon comes practically to an end. If the liquid is then poured carefully away from the filter-paper, and if a small portion of the latter is examined with the microscope, it will be found coated with the above mentioned filaments of textile matters, which effectually choke up its pores. If we attempt to filter unprepared sewage through coke, gravel, sand, peat, asbestos, or ordinary arable soil, we encounter the same difficulty, the rapid flow observed at first becoming sooner or later obstructed. Hence, as I shall endeavour to explain more fully below, filtration is not practically successful even for freeing sewage from its suspended impurities.

There is another constituent of sewage harmless in itself, but which seriously interferes with most processes of treatment. I mean the sand, gravel, and pulverized stone which are washed into the sewers by every heavy shower, to the greatest extent in towns, where the streets are "metalled" or macadamized. We shall show elsewhere how this silt, by keeping bad company, acquires offensive properties. Its inconvenience in the various forms of sewage treatment will be duly noticed.

In a manufacturing town or district the sewage is of a much more complicated character, and is more unsightly, though not necessarily more dangerous to public health—often, indeed, less so. Its purification is sometimes easier than that of the sewage of residential towns.

The pollution reaches its greatest height in places where the textile, tinctorial, and chemical arts are carried on. Here we find in the sewage, in addition to the normal excrementitious matters, sulphuric, hydrochloric, and nitric acids; alkalies, soap-lyes, solutions of iron, zinc, tin, alum, copper, chrome, antimony, and arsenic; waste dye liquors, spent dye wares, glue, sizes, dressings, waste tan solutions, etc. It must not be supposed that all these substances will be found in the sewage of a manufacturing town at one and the same moment. Many of them, indeed, neutralise and precipitate each other—a circumstance on which is founded a simple process for dealing with liquid industrial refuse. But if we watch the flow of such sewage, we shall find striking changes both in its colour and its odour, according to the kind of waste just emitted from one or other manufacturing establishment. I have repeatedly seen some agent proposed for sewage purification fulfill all requirements for several hours in succession, but on the sudden discharge of a new kind of impurity into the sewage it has been found not merely useless, but injurious, actually intensifying the evil.

The sewage of such towns, though containing a larger proportion of solid matter, both suspended and dissolved, than that of a residential town, is of far lower agricultural value—a point to be had in remembrance in selecting a method for its purification. Some of its possible ingredients, indeed, such as salts of zinc, proto-salts of iron (ferrous salts), are not merely useless, but positively injurious to the land.

The sewage of districts engaged in the metallurgical arts, the manufacture of hardware, etc., contains little extra organic impurity. On the other hand, it often holds in solution iron in considerable amount, derived, *e.g.*, from "pickling" iron wire. Copper and tin are rarely present, the value of their solutions being a guarantee that they will not be knowingly or wilfully run into the sewer. But certain salts of iron, as ferrous chloride, formerly known as muriate of iron, produced whenever metallic iron is scoured or cleaned from rust by means of hydrochloric acid (muriatic acid, spirit of salt), are abundant in certain kinds of sewage, and are not merely hurtful to crops for the time being, but permanently deteriorate the soil.

The sketch of the sewage of manufacturing towns just given is, of course, very general. Each such town, in fact, turns out a special type of sewage, the nature, effects, and treatment of which can be understood only after careful observation and experiment.

In all towns there is generally a well-marked difference between the day and the night sewage. In a residential town the sewage, from midnight till five or six in the morning, is very much reduced in strength, as well as in quantity. Often it consists of little more than the surface-water from the streets and roofs, and of the ground springs, which find their way into the sewers to a considerable extent. As the day advances the flow of sewage becomes more copious and more offensive, and is at its worst from two to eight p.m. In manufacturing towns the difference between the day and the night sewage is often less marked, since certain kinds of waste waters are discharged in the night, if requisite, and refuse which should not rightfully be run into the sewers at all, is often introduced when there is least chance of detection. Hence it is a serious mistake to imagine that the night sewage may be safely neglected and be allowed to pass into the rivers unwatched and untreated.

On Sundays, in a residential town, the sewage differs little from its condition during the rest of the week. In manufacturing towns it is on Sundays more purely excremental in its character, the industrial waste waters being in great part absent. The same holds good with respect to public holidays. In small agricultural centres the sewage on market days is perceptibly stronger than on other days, and requires an extra share of attention.

The question has often been raised whether there is any marked difference between the sewage of a "closetted" town and one where a proportion, larger or smaller, of the inhabitants make use of the old-fashioned cesspools or of dry closets and kindred appliances. Certain authorities, on the faith of their analyses, have maintained that there is little or no difference in the strength of the sewage. Careful consideration will show that this is scarcely possible. It may be at once admitted that the liquid excretions of man, and of all analogous animals, contain a larger proportion of nitrogenous matter than do the solid excreta. But in a closetted town the whole of the excrements of the population, liquid and solid, find their way into the sewers. In an unclosetted town, not merely the solid, but the great bulk of the liquid excretions are discharged into the cesspools and dry closets. Very frequently also, in small, straggling towns, public urinals, if they exist at all, are not connected with the sewers. Thus the water running in the latter consists, to a great extent, of soap-suds, the washings of pots and pans, and, in short, what is technically known as "slops." I do not see therefore, how it is possible, all "dilution" notwithstanding, for the sewage of an unclosetted town to contain the same matters as those of a closetted town.

A very important consideration is the water supply, known and unknown. By the "known" supply is meant the average quantity per head of the population served out from the water-works. By the "unknown" is meant the quantity which leaks into the sewers from other sources, as well as the surface drainage. Where the sewerage system lies in a wet sub-soil, full of ground springs, and where the sewers themselves are not water-tight, the flow of sewage in dry weather may rise to one hundred gallons per head of the population per twenty-four hours; whilst in towns where the water supply is scanty, and the sewers well made, it may fall as low as twenty gallons.

Another capital distinction is the space covered by a town in respect to its population. Where the streets are wide, where there are few persons to a house, and where there are many gardens, paddocks, vacant spots of building ground, etc., the sewage, especially in wet weather, loses many of its distinguishing properties.

This brings us to the question of single or double sewerage. It is suggested by certain engineers that the water-closets, the domestic sinks, the public urinals, etc., should discharge their contents into one set of sewers, whilst the surface drainage is carried off by a distinct system. It must be admitted that this double arrangement has some striking advantages: the sewage, properly so-called, would be at all times much more nearly alike in strength and quantity, which would facilitate its treatment, whether by irrigation or precipitation; its value would also be much greater. On the other hand, the expense of double drainage is a very serious consideration. The water from the street gutters, though receiving nothing from the urinals and closets, will be far from clean, and its admission into the rivers, despite the catch-word " the sewage to the land, and the water to the river," will be a very doubtful question. In manufacturing towns there are waste liquids—such as the rinsing waters from dye-works—which, if run into the sewage, will much dilute it ; but if allowed to pass into the rivers will greatly spoil their appearance.

Lastly, it is sometimes found, even where the common single system prevails, that the sewers require flushing to prevent solid matter from being deposited in parts where the gradient is insufficient. It is reasonable to suppose that this will be much more frequently the case in a sewer which carries fæcal matters only, undiluted with surface water. Here, then, we should get back, occasionally at least, to that dilution of the sewage which was to be avoided by the double system.

CHAPTER II.

HOW, WHEN, AND WHERE IS SEWAGE INJURIOUS?

BEFORE entering upon these questions I may, perhaps, be asked what is here meant by "injury," or "nuisance"? In my opinion these terms may fairly be taken in a wider sense than it is usually done. Judge McCarter, in trying the Newark case of river pollution, ruled that it was not necessary for the offensive matter complained of to be present in actually poisonous quantity in order to constitute a punishable nuisance. "It is sufficient if it render the water offensive or disagreeable to the taste or smell." Whether this decision be sound law I cannot presume to say, but it is assuredly sound common sense. It is not enough to contend that some particular matter poured into the streams, allowed to diffuse itself into the air, or to soak into the ground, does not directly and palpably cause some definite form of disease, or formally prevent the successful carrying on of some trade or calling. If it occasions discomfort to persons living near—if it offends their senses of sight or smell—the public have surely a full right to complain, and to demand judicial or, if necessary, legislative interference. My definition of "nuisance" is, therefore, wider than that too commonly accepted. I should include under this head any matter, whether solid, liquid, or gaseous, which is in itself injurious to health, or which may become so in contact with other substances, whether the latter may be in themselves hurtful or not; further, any matter which, though not demonstrably poisonous, is offensive to the senses.

There are, of course, certain limitations to be kept in view. Solid, non-volatile matter, on private premises, cannot be regarded as a nuisance so long as it can neither contaminate the air, nor be washed into watercourses, nor out on to public roads, or the lands of other persons.

Sewage contains, or rather consists of, in a large proportion liquid, or, at least, soluble matters, which, being liable to rapid chemical changes, give off volatile products—vapours and gases—in abundance. These vapours and gases are highly offensive to the sense of smell, and, if not directly poisonous, as it is still often maintained, they lower the vital tone of persons who habitually breathe air with which they are mixed. The greatest danger of sewage, and of water to which it has been added, is that it generally, if not always, contains minute living beings, bacteria, bacilli, etc., some of which are found to be casually connected with infectious diseases. These tiny organisms, named collectively microbia, or micro-organisms, are liable to increase and multiply in the water of rivers, wells, or pools into which sewage finds its way. Such water is thus rendered unfit for consumption by human beings, probably also by cattle. It may occasion sickness and death if used; *e.g.*, for rinsing out milk-pails, washing cooking utensils, not to speak of watering milk.

It may be urged, in opposition, that there are places, many of which I know, where the only water available for domestic consumption is drawn from shallow wells, separated from cesspools only by a few yards of chalk, gravel, or other open, porous subsoil; yet the general standard of health in the district remains good.

To this it must be replied that the residents of such neighbourhoods have, by a process of "natural selection," become inured to the effects of polluted waters; whilst a stranger coming to live in such localities is often seriously affected. But, above all, the health of a population using polluted waters depends on what is commonly called accident. If a simple case of typhoid fever, dysentery, or cholera, is introduced into such a district, the disease spreads on all sides, and commits sad havoc. Thus persons who drink impure waters hold their health or their lives at the mercy of chance.

But the microbia of contaminated waters find their way also into the air, and may be inhaled with it. In every highly polluted river fermentation is constantly going on. If anyone carefully watches the Aire at Leeds, the Kelvin Water at Glasgow, the Irwell, Irk, and Medlock at Manchester, he will see, especially if the day is warm and the barometer low, bubbles rising to the surface and bursting. These bubbles contain sewage gas, a mixture of compounds of carbon, hydrogen, and, to a less extent, of sulphur. In bursting they carry up with them the microbia, or disease germs, above mentioned, which thus become diffused in the air; whether these microbia are also carried up into the air when polluted waters evaporate quietly, without the escape of bubbles, has been disputed, but it may now be regarded as experimentally demonstrated.

There are, of course, cases where the volume of the sewage poured into a river may be very trifling compared with that of the river itself. Even London itself could not suffice to pollute the Amazon. But most of the great cities of the modern world discharge their refuse into comparatively small streams.

Seeing, then, that sewage is in the wrong place if poured into waters, we have to ask how does it behave on the land? Better, perhaps, but not quite free from reproach. It is then spread out over a larger surface than when it flows in a sewer or in a river, and it necessarily exposes a larger surface to evaporation in any one locality. However porous the soil, and however complete the drainage, a quantity of the evil-smelling liquid rises up into the air, carrying with it disease germs, if such be present. It will be found that if water is allowed to flow, however gently, over the surface of dry soil, bubbles are formed and burst. Thus, whenever sewage is turned on to land which has not been kept constantly damp, disease germs will be carried into the air, just as in the case of the bubbles which form on a polluted river. Further, the putrescent matter and the disease germs will be absorbed by two-winged flies (diptera), and distributed over the food and the persons of human beings.

Summing up this part of the subject, it may be safely asserted that sewage is harmful and offensive by its odour and its appearance, and especially by its affording a pabulum and breeding ground for disease germs. It is offensive in the water-courses and rivers by rendering the water unfit, not merely for human consumption, but for all delicate manufacturing processes. This last point is often overlooked. It may seem a convenience to the manufacturer to discharge his waste waters into the river, but by so doing he renders it of little use, save as a sewer, to all establishments lower down stream.

The sewage even of a residential town, except in very small proportions, unfits a river for the use of bleach, dye, print, or colour works.

The question has been raised by a pseudonymous writer in *Ashore or Afloat* whether sewage is truly detrimental to fish. We must here remember that there are fish, and fish. Not all species are alike in the conditions under which they can flourish, or even subsist. But one point, at least, is analytically proved. The greater the proportion of organic pollution in a stream, the smaller is the percentage of free oxygen held in solution, such oxygen serving to oxidise—in other words, to burn up—the impurities, and proving insufficient. Again, we know that certain fish—*e.g.*, the trout—can live only in well aërated, highly oxygenised waters. Putting these two considerations together, we can have little hesitation in pronouncing sewage pollution to be at least one of the causes which have tended to reduce the fish in our rivers. That in many cases, such as the Thames and the Clyde, other causes are at work, especially steam navigation, is highly probable.

We have now to consider what are the compounds or principles which make sewage water unfit for domestic purposes—drinking, cooking, washing, bathing, etc., and for the use of cattle.

Foremost come the compounds of nitrogen. These are of four kinds. There is nitrogen in organic combination, spoken of by chemists as "organic nitrogen" or "albumenoid ammonia." It is contained in albumen, gelatine, and, in general, in all the complex liquid or semi-liquid bodies of animal origin. These substances are introduced into sewage in the shape of blood, urine, pus, mucus, half-digested animal food, as well as by certain vegetable products. All such substances pass very readily into intense putrefaction, and are not only exceedingly offensive, but serve as nutriment for those low forms of animal and vegetable life which have been already mentioned as especially dangerous.

The second state is urea, which forms a very large part of the solids held in solution in urine. Urea is not dangerous in itself, and, in contact with a ferment which is never absent in sewage, it is quickly resolved into ammonium carbonate. It may therefore be regarded as a mere transition compound.

The third and fourth states in which nitrogen occurs in sewage are as ammonia and as nitric and nitrous acids, the two latter, in combination either with the ammonia or with potash or soda, forming nitrates and nitrites.

Ammonia, with its salts, and the nitrates and nitrites, even in the largest proportions in which they are ever met with in sewage, are harmless in themselves, though, like organic nitrogen, they may afford nourishment for microbia. Moreover, these forms—organic nitrogen, ammonia, and the nitrates—are constantly passing and repassing into each other. Growing fungi, and other plants, convert ammonia and the nitrates into organic compounds of nitrogen. The only way to render any water absolutely incapable of nourishing low forms of life is to keep it free from combined nitrogen in every shape and state. But, with waters exposed to the air, this, in any absolute sense of the words, is impracticable.

Phosphoric acid is another ingredient introduced into

water by contamination with the excrements of animals, or with any other decomposing organic matter. It is no absolute proof of the presence of such pollution, since many natural waters, on careful examination, may be found to contain it in very small quantity.

It is, in itself, not merely harmless, but doubtless in most cases beneficial. Yet it may be an indirect source of danger by favouring the multiplication of microbia.

Much the same may be said of potash. It is noteworthy that the three constituents—combined nitrogen, phosphoric acid, and potash—which are most valuable on the land, as being necessary to the growth of our food-crops, should be most dangerous in the water as fostering the growth of disease germs. One and the same kind of matter, accordingly as it is in the right or the wrong place, becomes the source of life or death.

Common salt (sodium chloride) is more largely present in sewage than in ordinary natural waters. It is, as every one knows, harmless to human life, even in greater proportions than it is ever known to occur in sewage. In water from a pump in London, now disused, it has been found to the extent of 60 grains per gallon. But it tells a tale of animal pollution. If in water there is more than 1 grain of chlorine per gallon (=nearly 1¾ grain common salt), we have good *primâ facie* reason to suspect that sewage, or at least the blood or the urine of animals, must have found its way in quantity into the river or the well.

There are, of course, exceptions, where salt springs and beds of saliferous minerals occur in the district, or if aluminium, iron, tin, etc., chlorides have been introduced by industrial waste waters. These latter, however, rarely find their way into a stream unaccompanied by sewage in the stricter sense of the term.

Waters containing merely putrefying vegetable matter, such as the drainage of rice-fields, may be positively poisonous. Mr. L. d'Aguilar Jackson, C.E., observed, when in Venezuela, that the result of getting wet in some rivers there, without subsequently rubbing dry, is malignant fever.

Thus the absence of chlorine (common salt), save in very trifling proportions, may be accepted as proof that a stream is free from *animal* pollution but not necessarily that it is a safe drinking water.

Soluble compounds of sulphur, especially hydrogen sulphide (commonly known as sulphuretted hydrogen), ammonium sulphide, etc., are very generally present in sewage if at all stale, and are a main cause of its evil odour, though not necessarily of its worst effects.

We must here note that, though putrefaction is in certain of its phases accompanied by a disgusting smell, we must not venture on the inverse conclusion that the absence of such a smell is any proof of the absence of the products of putrefaction.

The sulphur compounds are in great part derived from the solid excrements of men and cattle, who feed largely upon cabbages, turnips, and other cruciferous vegetables, and from water in which such vegetables have been boiled.

Sulphuretted hydrogen is also found in abundance when organic matter undergoes decomposition in presence of sulphates, such as gypsum. This reaction, as we may remark in passing, is a reason against the use of gypsum (sulphate of lime, calcium sulphate) in any form in the treatment of sewage.

I once met with an instance of gypsum having been added to sewage mud to aid in solidifying it. Sulphuretted hydrogen was given off in such plenty that the men employed suffered from temporary blindness, a well-known effect of this offensive gas.

Volatile compounds of phosphorus (hydrogen phosphide, phosphuretted hydrogen) have been popularly supposed to be given off by sewage and sewage deposits. I know of no analytical evidence in proof of this notion, and consider it a pure freak of imagination.

Marsh gas (light carburetted hydrogen or methane) is found in abundance in stale sewage, as also in sewers with too small a gradient, in ill-managed subsidence tanks, and in accumulations of sewage mud where decomposition is not checked by proper chemical agents. If such deposits are stirred up, and if a light is held over them, the bubbles of gas ignite and burn with a very pale flame. Marsh gas has no especially injurious action on the animal system.

Carbonic acid (carbon dinoxide) is another gaseous product of the decomposition of sewage, and is given off mingled with methane. Workmen have been suffocated by this mixture in ill-ventilated sewers and other underground passages.

A gas invariably present in natural waters, but practically absent in sewage, and found only in exceptionally small proportions in polluted rivers, is free and uncombined oxygen. In sewage it is spent or consumed by acting upon the organic pollution. This negative feature of sewage and of impure streams accounts for a part of their injurious action upon the higher aquatic plants and upon animals.

As regards "hardness" (*i.e.*, the presence of the salts

of lime and magnesia), sewage differs little from the ordinary water supply of the district. If anything, it will be softer on account of the introduction of soap and washing soda, and of a considerable proportion of rain water.

The sewage of a residential town has generally an alkaline reaction; that of a manufacturing town may be at times slightly acid, owing to various kinds of industrial refuse.

It will be at once seen, on considering the extreme complexity of sewage and the unstable character of most of its components, that as it flows it must be continually changing its character; that a sample taken in the heart of a town will differ in its chemical properties and in its physiological action from another sample which has travelled along the sewers for the distance, say, of two miles. These changes are most decisive where, as in certain manufacturing towns, the sewage has a temperature of 60° to 70° Fahr. It is needless to say that organic matters suspended or dissolved in water at such temperatures must undergo very rapid fermentations. Wherever the sewers are open, as at the grids and ventilation holes, clouds of steam rise up, carrying with them a very sickening odour.

Here, I may remark, is a serious flaw in the water carriage of excrementitious matters as at present conducted. If there are no ventilators the sewage gas is liable, under certain very possible contingencies, to be forced back into the houses, overcoming the traps of the sinks and water-closets. If there are ventilators of the ordinary kind, *i.e.*, gratings or trap-doors along the streets, all the passers-by are forced to inhale the fumes. In some towns, *e.g.*, in many parts of London, these trap-doors are constructed in the foot ways, and are opened for a certain time daily to reduce the pressure within. At such times it is a common sight to see a small crowd collect round the opening, craning out their necks as if to lose no chance of inhaling the vile fumes.

So far I have been speaking of the ingredients of sewage found to be harmful to men or other animals living near them. But, as far as fish and other animals living in the waters are concerned, several other impurities found in sewage and in polluted rivers must not be forgotten. Foremost must come, perhaps, solutions of free chlorine and of the hypochlorites, such as bleaching lime and bleaching soda. When these liquids find their way into a river the fish are destroyed far and wide. The surface of the Medway, at Maidstone, is sometimes covered with their dead bodies to such an extent that an exceedingly offensive smell is given off. H. H. Saare and Schwab (*Archiv für Hygiene*, vol. III., part 1, page 81) have observed that liquids containing 0·04 to 0·005 per cent. are rapidly fatal to tench, while solutions of 0·0008 per cent. are directly deadly to trout. A change into fresh water did not restore fishes after they had fallen on one side. As the lowest limit of this destructive action may be taken a proportion of 0·0005 per cent. of chlorine acting for two and a half hours. The presence of acids increases the action of the chlorine.

Of sulphur compounds I have already made some mention. Sodium sulphide (sulphuret) is, like not a few other substances, the more harmful the higher the temperature. Tench could bear for half an hour the proportion of 0.1 per cent.; and for two hours twenty-six minutes the proportion of 0·05 per cent. Their natural colour was, to a great extent, removed by the experiment, and did not return even on prolonged sojourn in pure water. Sulphuretted hydrogen was regarded fatal in proportion of 0·01 and 0·001 per cent., and proved deadly also to tench.

Lime in its caustic state, whether introduced into the stream as quick-lime, slacked lime, or lime-water, is well known as a fish destroyer, and has for ages been commonly employed by poachers. Its application in the treatment of sewage is common, and the attendant dangers are too often overlooked even by professed experts. I should advise riparian owners, lessees of fisheries, etc., to protest against its introduction into their waters. One manner in which lime present in waters destroys fish is by entering their gills, and, being there precipitated by the carbonic acid exhaled, it forms deposits of carbonate of lime, which interfere with respiration.

Common salt, even in the proportion of 10 per cent., was found harmless at temperatures ranging from 43° to 68° Fahr. Ten per cent. of chloride of calcium (formerly known as muriate of lime, and by no means to be confounded with chloride of lime) is harmless at 43° Fahr., but becomes hurtful, and even deadly, about 68° Fahr. Sulphate of soda, accidentally escaping into a river in unknown quantity, has proved very widely destructive to fish.

Ammonia and ammonium carbonate are inactive in any proportion likely to be met with in rivers.

Sulphurous acid, especially if accompanied by another acid, is even more deadly than chlorine. But its normal salts are harmless, and salts of lime diminish its injurious action. Hence the introduction of sulphurous acid is less pernicious in hard than in soft waters. Carbonic acid at 0·1 per cent. kills in a few minutes, while at 0·075 per cent. it has no permanent action.

Hydrochloric, sulphuric, and nitric acids are injurious. The first-mentioned at 1 per cent. is invariably fatal both to trout and tench. Sulphuric acid at 0·1 per cent. is fatal to trout in two to six hours; whilst tench are not seriously affected in eighteen hours. From the duration of the resistance against these three acids, Saare and Schwab infer that the higher the molecular weight of an acid, the less rapid is its action. Oxalic acid at 0·1 per cent. had no action on a trout in thirty minutes. Tannic acid at 0·1 per cent. is harmless even to trout.

Soda at 1 per cent. is fatal to trout on prolonged exposure. Its occurrence in streams to this extent is exceedingly improbable.

At 46 degs. Fahr. a tench remained for twenty-two hours without injury in a 5 per cent. solution of manganese chloride, and a trout endured a 1 per cent. solution for five hours.

Iron is a specific poison for fishes, both as a ferrous (proto) and a ferric (per or sesqui) salt; 0·02 to 0·01 of ferric oxide in a solution is injurious. Ammonia and potash-alum have as acute an action as the salts of iron, the limit of endurance falling between 0·1 and 0·05 per cent. As the poisonous action is said to depend on the proportion of alumina, it is probable that the simple salts of aluminium will be poisonous also.

It must here be remembered that the proportion of aluminium and iron compounds used in the treatment of sewage can be, and practically is, much more completely regulated than that of lime. It is very rare to see an excess of a sulphate or chloride of aluminium passing out into a river.

Arsenious acid in the proportion of 0·1 per cent., whether free or combined with soda, is not poisonous to trout and tench.

Mercuric chloride (corrosive sublimate) is at once fatal in proportions of 0·1 to 0·05 per cent. Copper sulphate in 0·1 and 1.0 per cent. kills trout in a few minutes if they cannot escape into pure water. Potassium cyanide, 0·01 and 0·005 per cent. is also deadly. Ammonium sulpho-cyanide and potassium ferro-cyanide have no effect at 1 per cent.

Carbolic acid is poisonous to trout in proportions between 0·01 and 0·005 per cent.

H. H. Saare and Schwab, in summing up their results, declare every substance soluble in water to be more or less injurious to fishes. Proportions which do not induce acute disease will probably be found hurtful on more prolonged action, and will especially interfere with the multiplication of fish.

Many of the results of these authors stand in need of verification, and their experiments have not extended to various substances, such as salts of lead, zinc and tin, chromates, etc., which may easily find their way into industrial waste waters, and which have even been recommended for the treatment of sewage.

We have next to consider what constituents of sewage, if any, are hurtful to plants. In the recent sewage of a residential town there is nothing in the least hurtful to any of our cultivated plants, unless it is either supplied in too large a quantity, or that it is too strong. Very few plants can bear repeated waterings with undiluted urine.

But in the sewage of manufacturing towns there are abundance of constituents which destroy or injure trees, crops, etc., and have, further, even a sterilising effect upon the soil. As such may be mentioned waste bleaching liquors, most sulphur compounds, including the waste of the alkali manufacture, sulphuric and hydrochloric acids, solutions of alumina, iron, tin, lead, zinc, chrome, etc. Further, waters containing tannin, gallic acid, starch, glucose, and many coal-tar products.

Thus objection must be taken to the waste waters of paper-mills, bleach-, dye-, print-, and chemical works, electro-plating establishments, and works where wire is "pickled," where the manufacture of tin and tern plates is carried on. On the other hand, the waste waters of soaperies, glue and gelatine works, and some other establishments are not to be feared. It will be at once seen that the useful disposal of waters injurious to vegetation is an exceedingly difficult task, and that some of the best methods are at once excluded.

CHAPTER III.

THE DISPOSAL AND TREATMENT OF SEWAGE.

IN the primitive condition of mankind population was necessarily scanty, and, above all, it was nowhere congested into dense masses. Hence the offensive products necessarily resulting from animal life in whatsoever form occasioned no trouble. Sewage, as we understand it, was not produced, and consequently no question concerning its dangers and its treatment could arise. Both the liquid and the solid excreta were voided on the ground, and, never accumulating in large quantities, they were easily absorbed by the soil, or were in part consumed by certain insects. Even to the present day there are places in England, not to speak of less populous countries, where this system, if I may so call it, still prevails. So long as the population remains thin, and as no epidemic is present, there is no manifest harm. Epidemics, indeed, are the rarer and the less formidable in proportion as a country is sparsely peopled. But so soon as it becomes at all numerously or densely inhabited, excrementitious matter accumulates in the neighbourhood of dwellings and is liable to be washed by heavy rains into wells and other waters used for drinking. Hence in times of any common sickness serious consequences may ensue.

It is interesting to note how, in so ancient a document as the Pentateuch, it was especially enjoined upon the Israelites that excrements should not be left to lie upon the surface of the soil, but should be covered with earth (Deuteronomy xxiii. 13.) This command evidently implies a knowledge of the sanitary efficacy of the soil—possibly also of the danger of exposing fæcal matter to be washed into the rivers, or to be the pabulum of flies, which then settle upon human beings and their food, and thus propagate disease. The possession of such knowledge at so early a date is a very remarkable circumstance, if we suppose it to be a dictate of Egyptian civilization.

The next step, unavoidable as soon as large villages and towns took their rise, was the cess-pool. This was, and often still is, a mere pit in the soil, not in any way rendered water-tight to prevent the contents from soaking into the earth at the bottom or sides. Nor was it generally covered in so as to minimise the escape of offensive odours. In most places the contents were dug or scooped out from time to time, and conveyed into the gardens or the fields. Here, at any rate, was an attempt to restore to the soil what had been taken from it.

But this method is open to very serious objections. Unless the cess-pit is made water-tight with masonry or cement, the liquid portions of the excrement—its most valuable part—soaked into the earth and were lost. Worse than lost, they not uncommonly found, and still find, their way into chinks and crevices in the sub-soil, especially in chalky districts, or drained through porous formations (gravel, etc.) into wells and watercourses.

A further objection is that the emptying such cesspools is a very loathsome process, and under certain circumstances it may prove highly dangerous both to the workmen employed and to all persons living in the immediate neighbourhood.

To lessen this drawback it became, from a very early age, customary to add to the contents of the cess-pits matters which, it was supposed, might to some extent absorb and neutralise the evil odours. Among such substances the ashes of wood, peat, coal, etc., took a prominent place. The cess-pool became also the general receptacle for all the refuse of the household, some of which was far from tending to render it less offensive or less dangerous. Such is the village cesspool in our days, tolerable only where it is lined with slabs of slate or flag-stones well cemented together, so as to prevent infiltration into the sub-soil.

But in many districts, where room was plentiful and where the soil is, as commonly termed, "light," the cess-pools were *not* emptied periodically, or indeed at all.

When full, they were covered in, not by any means in an air-tight manner, whilst a fresh pit was excavated near at hand and brought into use instead. There are towns where this process has been carried on so long and so generally—though now put an end to—that the entire subsoil is polluted to an unknown depth, and its sanitation becomes scarcely possible. It is, or was until very lately, common to find in castles and other large mansions old cess-pools, the very existence of which had been quite forgotten. I understand that subsequent to the death of Prince Albert not a few such abominations were detected in Windsor Castle. The same evil was not wanting in dwellings of much less pretentions. It often happens in country places, when a row of old cottages is demolished to make room for some new buildings, the cess-pits are not cleared out, but simply filled up with any kind of available rubbish, and new houses or manufacturing premises are built over the spot, the occupants being in happy ignorance of what lies beneath their feet.

To sum up this part of the subject, it may be said that cess-pools are utterly out of the question in towns. They are permissible, nay, sometimes they may be the best expedient, in the case of villages and detached houses in the country. But the following conditions must be observed :—

1. The pit must be lined in a water-tight manner with masonry, flag-stones, etc., so that nothing can leak out into the soil.

2. It must be situate so that a clear current of air can play between it and any dwelling or workshop, etc., and it should be covered in from rain.

3. It must be regularly cleared out and the contents dug into the soil *not* in the immediate neighbourhood of any well or water-course.

4. Garden mould or fine ashes should be added to the fæcal matter liberally.

5. At the time of thus emptying out, it is well to add some active disinfectant. The like should be done if any infectious disease occurs among the persons making use of the convenience, or indeed in the neighbourhood. The emptying should, if possible, never be performed in warm, calm, moist weather. Frost, or, in its absence, drying winds afford the best opportunity.

We turn now to various improvements on the cess-pool system. Foremost among these stands the earth-closet or dry-closet, first proposed by Mr. Goux, and since variously modified. Its principle is that a quantity of dry earth, peat-mould, charcoal, etc., is stored up behind and above the seat of a closet much resembling that of an ordinary water-closet. After this has been used a handle is pulled, and a quantity of the dry material just mentioned falls down so as to cover the excreta and prevent the escape of any offensive fumes. Daily or oftener, the box, pail, or other receptacle, is emptied out either by the inmates of the house or by public servants.

This system has very great advantages from a sanitary point of view. There is, with ordinary care, no nuisance, no danger to health, no generation and escape of sewer gas either in-doors or in the streets and roads; no pollution of rivers, whilst the liquid and solid excrements are restored to the soil undiluted, and before they have had time to enter into fermentation.

But, on the other hand, there is considerable trouble and expense in preparing the absorbent material, which must be dry, conveying it into the closet, and emptying the pail or receiving-box. This system, too, like the water-closet scheme, fails to make provision for vegetable refuse, and not only so, but for soap-suds and washing waters, which have still to flow into the sewers, and may easily be the carriers of disease germs.

In some cases the dry absorbent matter is omitted, and the excrements are received into tubs or cans capable of being closed air-tight after use. This system has been adopted in some small towns, where carts go round daily to receive the full pails and leave empty ones in their stead. It is, of course, essential that these vessels, of whatever shape and material, should be thoroughly cleaned when emptied, or the nuisance would be serious. Hence the working cost is considerable. Thus, though this system has given satisfaction in some small towns, it is out of the question in larger cities, where the distances to be traversed would require the employment of a ruinous number of carts, horses, and men.

In another form, there is in each house a fixed receptacle into which all excreta are voided, and which is fitted after use with a (supposed) air-tight cover. From the bottom of this receptacle runs a pipe, which passes through the outer wall of the house, and is closed at its end with a plug. In the night, carts go round to each house. The plug is withdrawn, and a pump is attached to the pipe, which draws the contents of the receptacle into the cart. This operation is not unattended with nuisance, and it is practically impossible for the receptacle ever to be thoroughly cleansed from putrescent fæcal matter.

This is, perhaps, the place to mention the system of Liernur who, unlike prophets, finds honour in his native country, if nowhere else. Here, also, there is no water of dilution. The movable pails and the collecting carts are suppressed, and the excreta are aspirated by powerful pneumatic machinery out of the recipients through a system of pipes. That either the recipient or the interior of the pipes can be entirely sucked clean in this manner, even if a perfect vacuum could be instantly created at the outer extremity of the system, is not to be expected or believed.

We thus come to the end of the systems which do not call in the aid of water, but which deal with the excreta of human beings as a whole, not first forming sewage, in the conventional sense of the word, and then striving to extract some of its constituents for useful purposes.

All these dry systems, except the primitive cess-pool, require the dust-bin as an accessory. All of them, as I have briefly hinted, fail to provide for household "slops." Still more completely they are out of the question for industrial waste waters, and even for the washings of slaughter-houses, and the drainage of piggeries and stables.

Hence in all towns a system of sewage becomes essential, and the notion of disposing of human excreta also in this manner naturally suggests itself. To this we therefore proceed.

CHAPTER IV.

WATER CARRIAGE IN GENERAL.

THE use of rivers, arms of the sea, and canals as the recipients of fæcal matter is not of modern origin. Certain antiquarians consider that the water-closet, as we now have it, was known in ancient Rome, and even that the "summer-parlour" of Eglon, King of Moab (Judges iii. 20), was a convenience of this kind. But in modern times it has been common, wherever dwelling-houses or factories stood on the brink of a stream, to have out-shot closets jutting over it, so that the fæcal matters might fall at once into the water. The margins of the Fleet Ditch, in London, so long as it remained open, are said to have been lined with such arrangements. In the East of Germany they existed about the year 1830. In the North of England they were not uncommon twenty years ago, even on the brink of rivers which were nearly dry in summer, thus occasioning a grievous nuisance. I have seen in a Midland district a contrivance, if possible, still more offensive. A row of cottages stood on a high ground, separated by a public road from a brook, or rather, ditch. Each cottage had a kind of cess-pool, from the bottom of which there ran an ill-made drain under the road. Whenever there was a fall of rain part of the contents of these cess-pools were washed through the drains, down the bank, and into the tiny brook, which was thus converted into an open sewer.

In Melbourne, with an average temperature higher by 10° Fahr. than that of London, cess-pools in houses were allowed to overflow into the surface-water gutters at the sides of the streets.

The first systematic use of the water-closet and of the water-carriage of fæcal matters began with the suppression of cess-pools in London and other large cities. The excreta of the population were thus compulsorily and by authority diverted into the rivers, no attempt at purification being thought of. Before proceeding to describe the various modifications of, and improvements on, this crude and nasty procedure, I must point out certain initial disadvantages which attend water-carriage, however the sewage matters may be ultimately disposed of.

There is, in the first place, a very serious consumption of water, over and above what would otherwise be necessary for domestic purposes. This increase may, perhaps, be roughly taken as 20 per cent. Thus if a residential town without water-closets requires 100,000 gallons of water daily, if these conveniences are generally introduced the daily demand will rise to 120,000 gallons. This increase may easily mean more than an addition of 20 per cent. to the cost of the water-supply. It is recognised on all hands that the water furnished to a town must be good in quality. But such water is not everywhere to be obtained in sufficient quantity within a reasonable distance. It might happen that a source fully adequate for a town in other respects might not leave a sufficient margin for the water carriage of excreta.

It may be contended that a second-rate water is quite good enough for working water-closets and flushing sewers. So it is; but we are then driven to the troublesome and costly expedient of a double water supply—a good quality for drinking, cooking, etc., and an inferior sort for the closets. This would prove a heavy addition to local taxation—a sphere where no one thinks of retrenchment.

Another disadvantage is the generation of "sewer gas." This term includes all the volatile products—gases or vapours—arising from excreta in various stages of decomposition. Such sewage gas may or may not be capable of directly occasioning disease. But few—very few—competent medical authorities will deny that if persistently inhaled it lowers the tone of the constitution, and renders the inroads of fevers, etc., more probable. Above all, this same sewage gas is liable to convey disease germs into our dwellings and into our bodies. It is a serious consideration that we have now this gas "laid on," so to speak, in every house. Not only our water-closets, but our bath-rooms and our very sleeping apartments, are placed, potentially at least, in connection with the sewer. If there is a chink in the piping, a defect in the soldering, or if the valves are out of order, this potential connection becomes actual. Hence, in addition to good plumbing at the beginning and careful management and watching afterwards, the water-closet requires precautions in its *position*. It may be a luxury or a convenience to have water-closets in the interior of a house, separated, perhaps, from the bed rooms merely by ill-fitting doors. It is very agreeable, doubtless, to have the wash-stand in a dressing-room fitted with a fixed basin from which the water can be discharged into the sewer by lifting a plug. But this convenience has to be paid for in the shape of risk. The true situation of the water-closet in a house is in an out-shot annex with two windows on opposite sides, so that it may be ventilated not into the house, but into the air outside.

It must also be remembered that water forms no absolute barrier between the air of the sewer and the air inside the house. Through the water there is a constant, though slow, interchange of gases going on.

As for the plugged wash-basins in bed-rooms and dressing-rooms, despite their convenience, they had better be given up. This change, I understand, has been carried out in many of the principal hotels in America.

Another drawback on the water carriage system is that the sewers, if not absolutely water-tight—which is scarcely possible—allow more or less of their contents to ooze out and saturate the surrounding soil, thus gradually forming a bed of poisonous matter. This bed alters the character of the "ground-water," which has been so thoroughly investigated by Professor von Pettenkofer, of Munich, and renders it more harmful. The vapours given off from this polluted "ground-water" will escape in the direction of least resistance, and that direction, in the case of streets paved, flagged, and especially asphalted, will be into the cellars or the sunk ground-floors of the adjoining houses.

If a momentary digression is permissible, I would add my protest against the custom, all but universal, of carrying the soil-pipes of water-closets under the floor of the house to join the sewer outside in the street. If the connections are imperfect, or have never been made at all—as is sometimes the case—ill-health and death will be the lot of the occupiers of the house. A sewer should never be permitted to pass under any human habitation unless there is a clear, open air-way left between the top of such sewer and the foundations of the house.

Another point is the ventilation of the sewers—an absolute necessity. If we suppose a system of sewerage perfectly air-tight, and having no inlets, save the trapped openings through which the liquids from sinks, water-closets, etc., find entrance, it would very often happen, from the fermentation of the fæcal matters, that the pressure inside the sewer would be greater than the pressure of the external atmosphere. In such cases the sewer gas would force its way through the ordinary sink-traps, and escape in bubbles into the houses. To prevent such a grave inconvenience it is usual in most towns to provide the sewers with trap-doors, which, when opened, act as ventilators, and allow the condensed gases, if any, to come to an equilibrium. But these ventilating trap-doors are placed too often in the foot-ways, with the bad results which I have already pointed out. Surely situations might be found for ventilating traps not open to this objection.

It will be perceived that the sewer retains its unpleasant character up to the very point where it discharges its contents for treatment. No system of purification adopted at this point can obviate the nuisance arising from every grid on the way. Disinfectants of various kinds are sometimes poured into the sewers in an intermittent manner, in the hope of preventing the formation of sewer gas. It sometimes happens that conterminous sanitary authorities use substances which are mutually incompatible, one, for instance, employing chloride of lime or a permanganate, whilst his neighbour is using some de-oxidising agent, such as carbolic acid or an alkaline sulphite.

I can here merely refer to two projects for dealing with sewage gases and the sewer ventilation question. The late Peter Spence, of the Pendleton Alum Works, proposed to connect all chimneys, whether of houses or factories, with the sewers, whilst these, again, were made to vent into one huge central stalk, actuated by a mighty furnace. In this manner all the sewage-gases would be aspirated into the central chimney, and, being already to some extent disinfected by the smoke and other products of combustion drawn into them, would be finally dealt with in passing through the furnace and discharged in a harmless state from the central chimney. Thus smoke and sewage gas would be done away with at once.

Mr. W. A. Gibbs, of Gillwell Park, Essex, proposes a different scheme for towns situate on the banks of tidal rivers. Huge fans, worked by tidal power, would be erected at the outfalls,—say at Barking Creek and Crossness. The air in the streets of London would rush *into* the sewers, removing fog, sewage gas, and all evil vapours. Whether it would be possible by any power applied at the outfalls to create an in-draught into the sewers at the most remote parts of the metropolis, I am not prepared to decide. But I could wish that both of these schemes might have a fair trial in some town, smaller than London, and conveniently situate.

Not the smallest of the defects of the water carriage of excreta is that the excessive dilution reduces their agricultural value, and renders any method of treatment that may be adopted less remunerative, if not less easy.

Lastly, we must remember that where water carriage

is adopted the dust-bin is indispensable. The solid refuse, household ashes, the bones, skins and shells of fish, the parings and stalks of fruit and vegetables, etc., cannot be passed down the soil-pipe, and hence they find a temporary home in the dust-bin. This receptacle is supposed to be from time to time cleared out by the servants of the local authority, or by contractors—an unpleasant operation, in which smells, offensive, if not positively harmful, are diffused. The nuisance is the greater the longer the refuse is allowed to accumulate. Such time may be somewhat long, especially if the householder is not willing to give the men employed a liberal backsheesh. But the worst evil is the subsequent destination of the matter. It is sold by the contractors to building speculators, who use it for all manner of unsanitary purposes, such as filling up hollows, laying out streets, mixing mortar, etc. I have even seen it employed in mending old-established streets, long since in the hands of the local authorities. The unpleasant smells, and the flocks of blow-flies hovering round, proved but too plainly that the "dust" was utterly unfit to be used in street repairing or in building operations. The contractor is supremely indifferent about complaints inserted in the Press, or addressed to the local sanitary department. He has duly "squared" the surveyor, who will, as in duty bound, "make it all right." But in spite of its shortcomings, and of its needful adjunct the dust-bin, the water carriage system is in all large cities simply Hobson's choice. No one has yet suggested any practicable method by which it might be superseded. Our inventors are too busily employed in devising "ærostats," submarine boats, and other devices for taking human life, to turn their attention in this direction.

CHAPTER V.

THE BAZALGETTE SYSTEM.

CONCERNING the direct discharge of sewage and other waste waters into rivers, nothing further need be said. It has no advocates, and offers no advantages save a fallacious cheapness. The whole vexed question of river pollution is the outcome of this mistake.

The first method that we shall discuss is very simple, very costly, and very unsatisfactory. The original responsibility for this process seems to fall upon F. Lipscombe, who proposed it in a patent, A.D. 1857, No. 2168. E. Strangman (A.D. 1861, No. 1040) obtained a patent for a somewhat similar scheme.

It may, however, be conveniently called Bazalgettism, from the distinguished engineer to the Metropolitan Board of Works, who has carried it out on a gigantic scale as a means of disposing of the sewage of London. Its essential principle is discharge either directly into an arm of the sea, or into a tidal river, at the time of ebb.

In the latter case, the sewage must be received, in the first place, in large reservoirs or store-tanks, in which it is allowed to accumulate during the flow of the tide, and is let out as soon as the ebb commences. The theory of the process is that before the tide again flows, all the

THE BAZALGETTE SYSTEM.

offensive matter will have passed out at the mouth of the river and have been diffused into the sea.

It is manifest that this method is available only in a limited class of places. If a city lies on the banks of a river, so far from the sea that the sewage would not have time to reach its mouth whilst the tide is running out, it is met by the next flood tide and carried back again. Thus the lower part of the river becomes a sink of pollution, intensified by constant recruits of filth.

Even on the sea-shore this system cannot be safely applied in the case of land-locked harbours, and narrow deep bays. In few situations is sewage pollution more formidable than in the harbours of Marseilles, Bombay, Rio Janeiro, Bahia, etc. Those of Sydney and Melbourne will, doubtless, in due time reach the same undesirable condition.

The expense of Bazalgettism is exceedingly serious. The construction of the tanks to store up the sewage whilst the tide is running up involves an outlay which for many communities is simply prohibitive. Such tanks cannot, for obvious reasons, be situate in or very near to the city or town in question. They must lie at a lower level than such city, or the sewage cannot be conveyed into them by gravitation. Yet they must be so high that their contents can be run out to the very bottom at low water. These two conditions will generally be found mutually incompatible. Thus, at the two London sewage outfalls, Barking Creek and Crossness, the sewage is first pumped up into the storage tanks. I have seen a calculation what every stroke of the powerful and ornate machinery effecting this task extracts out of the pockets of the ratepayers.

It will be at once seen that this system offers not the

slightest set-off against the expenditure. It is all outlay, with no returns.

Nor is the prevention of nuisance by any means complete. I will suppose, for argument's sake, that every drop of the discharged sewage is carried out to sea before it can be forced back by the returning flood tide. Yet the lower part of the river is of necessity polluted to the annoyance and danger of persons navigating it or dwelling near its banks. A portion of the finely divided mineral matter which invariably enters into the sewage of a town, especially where the lamentable custom of macadamising or "metalling" the roads is tolerated, is gradually deposited in the bed of the stream, and *must*, all assertion to the contrary notwithstanding, take rank as one of the agencies which silt up the channel. Each such particle of sand or clay or ground-stone becomes, by surface attraction, coated with a thin layer of the offensive organic matters present in the sewage. Any one who doubts this may soon be convinced to the contrary if he will take a pound of such sewage silt, dry it, and then heat it strongly, when a powerful fæcal odour will be given off.

Even in those few localities where the outflow can take place directly into the sea, there is still an opening for mischief. The suspended particles of excrementitious matter "hug the shore," just as corks or chips floating in a tub of water make their way by attraction to the side. Among the unpleasant results of this phenomenon I may mention that shrimps feed upon the unclean matter, and are then caught and consumed by men before it can be completely assimilated. It has been lately established that mussels found in the open sea are harmless, but if deposited in harbours they become poisonous, and lose anew their malignant properties on being taken back to the open. It is exceedingly probable that the poisonous properties developed are due to the consumption of sewage matters. Two of the poisons detected in mussels by Dr. Brieger are ptomaines.

Where sewers open into the sea without the interposition of a storage tank, the case is still worse. When the tide rises and covers the mouth of the drain pipes the sewage gas, holding possibly disease germs in suspension, is driven back into the town, and into the very houses, overcoming the traps.

In the rare cases where a manufacturing establishment is situate on an open coast—not a tide-way or harbour—its liquid refuse may find vent direct into the sea, if no more useful mode of disposal can be devised.

But the admission which we made above, for argument's sake, is found, in the case of London, to be quite incorrect. Sewage matters discharged into the river at Barking and Crossness are not pushed out to sea by the combined action of the ebbing tide and of the current. They mingle with the water, and work their way back to points far above the outfalls, thus effecting that pollution which the intercepting sewers and the costly channels running parallel to the river were to have averted.

The Bazalgette process, as applied to London, is a total failure. It involves the utter waste of all the manurial matters in the sewage; it aids in silting up the bed of the Thames; it occasions a nuisance much complained of by the inhabitants of the country below the outfalls on both banks; its cost is exceedingly serious; and it does not even guarantee to the inhabitants of London an unpolluted river. Some persons might

SEWAGE TREATMENT.

think that all these charges against the system being admitted, the Metropolitan Board of Works would hasten to abandon the scheme altogether, and cry *peccavimus* with the best grace possible. Whoever should form such an expectation would betray a very imperfect knowledge of the official mind. Instead of renouncing the whole scheme as a failure, they seek its extension. It is proposed to convey the sewage down to Thames Haven in a vaulted culvert, that from the south side of the Thames being carried under the river by a tunnel and then pumped up into the culvert on the northern shore. The cost of this gigantic sewer, twenty miles in length, is estimated by Sir Joseph Bazalgette at £4,000,000. Those who have had some experience in engineering estimates will expect this sum to be largely exceeded should the scheme unfortunately be carried into execution. The working cost with which the ratepayers of the metropolis are threatened is £317,000 yearly, in addition to interest on the four millions borrowed and the expense of paying off the loan.

At Thames Haven the entire sewage will be discharged.

We will now examine the probable result of this scheme. The annual charges for dealing with the sewage of London, already quite heavy enough, will be seriously increased. The fæcal matters of the sewage will still, as at present, be wasted. The dwellers on the shore near the point of outflow will complain of a nuisance, just as now do their neighbours a little higher up stream. Sewage matters will scarcely find their way, as at present, to points above London Bridge, and the silting up process will be suspended. But, on the other

THE BAZALGETTE SYSTEM.

hand, the Thames will be deprived of a tribute of 120 million gallons of water which are now poured into it daily, at a time when its level is lowest owing to the ebb. Surely it would be far better to pour this volume of water into the river in a purified condition, which, as we shall see, might be effected more cheaply, and with at least a partial utilisation of the manurial matters contained in the sewage.

It is well known on the west coast of Africa and in Mauritius that the contact of polluted land waters with the sea is greatly to be dreaded, as far as public health is concerned.

CHAPTER VI.

IRRIGATION.

THE second method of sewage treatment is irrigation; a system which has contrived to secure advocates, many, influential and enthusiastic. Its essential principle is the application of the sewage to cultivated land, under the impression that the excrementitious and other offensive and putrescent matters held in solution or suspension will be absorbed by the soil, and utilised by the growing crops; whilst the water, sufficiently purified, will pass away into the drains, and may be permitted to enter the rivers without fear of pollution.

The action is supposed to be two-fold; on the one hand, the soil itself has an absorbent power. As the sewage soaks down into the ground, it carries with it air, by which certain of its organic impurities are oxidised and destroyed. That such a purifying process really takes place, whatever its theory, may be proved by the simple experiment of filling a flower-pot with soil, and pouring a little sewage, etc., upon it. The water which oozes out below will, unless the quantity is too great, be to a very considerable extent deprived of its offensive smell.

On the other hand, the growing vegetation is supposed to absorb and assimilate a proportion of the organic matter present, just as it does any other manure. The mere filtering action of the soil is thus reinforced, and the impurities are not merely removed, but turned to account.

Hence, irrigation is, under favourable circumstances, very efficacious. It requires, however, certain conditions, which are not everywhere to be met with in combination. The first essential is, of course, a large plot of land, which should be at some distance from the town supplying the sewage, and should not border upon any populous district. The irrigation farm, as such a plot is called, should also be situate at a lower level than the town, as, if the sewage has to be raised to a higher level by pumping, both the first outlay and the annual working cost are necessarily and heavily increased.

Now, as nearly all our inland towns lie on the banks of some river, the only ground to which the sewage can flow by gravitation will generally be found at some point lower down the river valley. But in the more populous parts of England and Scotland the river valleys are generally occupied by a string of towns, villages, and factories. Hence, land in such situations is very valuable, and to acquire a plot suitable for an irrigation farm, if practicable at all, is, necessarily, a costly undertaking. When such a plot has been found, a further outlay is incurred in laying it out, and in conveying the sewage thither. The latter point may be very serious: the authorities of a town in Lancashire, having conveyed their sewage to a plot of land at a certain distance, were sued for heavy damages by the owner of some subjacent coal deposits, on the plea that the working of such deposits would be interfered with by the sewage channel.

Another consideration is that if the sewage has to be conveyed to some distance before it can be applied to the land, the river of the district must be, for a part of the course at least, robbed of no small share of the waters which it would otherwise receive. This point is the more serious, as many streams in the manufacturing districts are very low in a dry season.

Again, the quality of the soil is no less important than the situation. If it is too open and porous in its texture the sewage passes rapidly through it, and emerges, scarcely, if at all, purified. This is especially the case with shallow, sandy, or gravelly soils overlying rocks. If, on the other hand, it is too compact and retentive, the water cannot escape readily enough, and the soil is rendered swampy. Perhaps the worst case is when the soil, as is not uncommon in chalky districts, is intersected by cracks or crevices which extend far away underneath the surface. Into these the sewage may pass unpurified, and may thus find its way into wells, water-courses, etc., at a very considerable distance, thus leading to unsuspected mischief.

It must next be remembered that not all kinds of sewage are fit and proper to be applied to cultivated land. If there is a notable proportion of industrial refuse, such as acids, metallic salts, dye and tan liquors, etc., it will injure or even destroy the crops, and may sterilize the land for a considerable time.

Another point too often overlooked is that a given plot of land cannot go on for ever deodorising and disinfecting an unlimited supply of putrescent or putrescible organic matter. On referring to the chapter on sewage it will be seen that certain of its constituents gradually choke' up any porous material, as was shown by an easy experiment. In addition, the soil, as in the case of an old and crowded graveyard becomes ultimately saturated, and would have to be left unused for a considerable time—probably many years—before it could again become capable of disinfecting putrescent matter. Hence in selecting a plot for an irrigation farm the possibility of extension, if needful, has to be taken into account.

I must here point out an erroneous assumption in an official document which appeared a few years ago. In speaking of the various towns which have adopted the irrigation process as a means of disposing of their sewage, and of the sums expended and the rates per pound required for paying off the first outlay and defraying the working expenses, it is invariably added, "the cost of acquiring the land will be paid off in so and so many years, and the rate per pound will then be reduced to such-and-such a figure," or words to that effect. But this flattering prospect is based entirely upon the assumption that a given plot of soil will go on for ever absorbing and disinfecting the excretions of an increasing population. This assumption is so contrary to experience in all analogous cases—*e.g.*, that of graveyards—that until it shall have been demonstrated by actual observation under the strictest test conditions—it cannot for a moment be entertained.

The next point is that of climate. In some countries the rain fall is too small, either positively, or in relation to the degree of evaporation from the surface of the ground. Or, if not insufficient, it is intermittent, and falls in torrents after long intervals of drought. In all such climates, especially if their average yearly temperature, or even that during the season when vegetation is active, is high, irrigation is the one thing necessary to convert a desert into a garden. Instances may be seen in Italy, Spain, Algeria, Syria, Persia, India, etc. Hence we often find Indian officers enthusiasts in irrigation.

But, on calm consideration, we shall find the climate of England differing from that of the countries above mentioned in almost every essential particular. Our average yearly temperature is only about 49 deg. Fahr., and even this small allowance of heat is made up rather by the usual absence of severe frost in winter, than by solar activity in summer. Our rainfall though absolutely small as compared with that of some countries, is large if we take into consideration our low temperature, our cloudy skies, and frequent fogs, which intercept the direct rays of the sun, and thus check evaporation. Besides, with us, rain descends not in few and violent gushes, but in a very frequent drizzle. So that, in short, with us, moisture is not the one thing needed for agriculture, but the great enemy of our farmers.

In a document which appeared some years ago, under the auspices of the Local Government Board, it was admitted that sewage irrigation is not well adapted to white crops and to potatoes, and that even to the ordinary root crops it must be applied with caution.

The fact is that, as was brought into distinct prominence by H. Lefeldt, a Prussian engineer sent over by his government to examine and report on the various systems of sewage treatment in use in England, sewage can be (to a certain extent) purified by passing through the soil, and it can also serve as a manure. But it cannot, under ordinary conditions, combine these two functions. The farmer could, even in England, often use sewage to advantage if he could have it now and again for a few hours in unusually dry weather. But if the sewage of a town is to be purified in this manner, it must flow upon the land day and night, summer and winter, seed-time and harvest, in wet weather and in dry. Nay, just when rain is most plentiful, and is consequently least needed, the sewers pour out an increased volume, which must be admitted upon the land. Hence the interests of the farmer and of the sanitary authority are in full opposition.

One plant, and, I believe, only one—Italian rye-grass —answers the conditions of both parties. Its appetite for moisture seems insatiable. Hitherto its cultivation has been checked by the great practical difficulty of converting it into hay, portable, and capable of being reserved for sale when the market was favourable.

At Whitley Manor Farm, near Reading, the rye-grass, the produce of ninety-four acres of land irrigated with sewage, has been successfully converted into hay by means of the "Harvest Saver" of Mr. W. A. Gibbs, of Gillwell Park, Essex.

This is a very decided advantage for irrigation, and, where that system is otherwise practicable, may render the treatment of sewage self-supporting, even if not positively remunerative!

As regards other crops, the benefits of irrigation are very doubtful from an agricultural or a horticultural point of view. At the celebrated irrigation farm of Gennevilliers, near Paris, where, by the way, all the circumstances are exceptionally favourable, the production of vegetables and strawberries is very large. But

E

there is this important drawback—these crops come in late in the season, when the market is already glutted, and fetch in consequence the lowest prices. The object of the farmer and the gardener being to bring forward their produce as early as possible, sewage farming is, at least in cold climates, like that of Britain, placed at a disadvantage.

The question may be raised, "Why should irrigation, or the application of liquid manure retard a crop?" I reply, because it lowers the temperature of the soil and of the stratum of air immediately over the soil. Water cannot evaporate without abstracting a portion of heat from any neighbouring bodies. Irrigation, therefore, chills the soil just as a wet coat chills the wearer, or a damp house the inhabitants. But if irrigation is kept up, day by day, the effect is the same as if the field were altogether removed into a colder climate. Few persons will, I think, admit that the soils of England are likely to be benefited by chilling.

The next question must be, What is the influence of continuous irrigation upon the texture and condition of a soil? Have we improvement or deterioration?

The action exerted is twofold. Suppose a quantity of porous soil or of potting-mould is placed in a flower-pot, filled with every appliance to facilitate drainage, and is then continually drenched with pure water. It will soon be found that the soil loses its open, porous character, and becomes compact and sodden, and any ordinary plants which may have been growing therein begin to show marks of declining health. In a sewage irrigation field the same process, of course, takes place. If, for any reason, the supply of water is interrupted, as, *e.g.* whilst gathering in the crops, the soil bakes together into a hard mass, which is afterwards not easily made permeable to air and water. This result may be very well shown in the flower-pot above-mentioned. If the surface, after receiving for some time a superfluity of water is allowed to go dry, such soil becomes most intractable.

But when soil is constantly drenched, not with pure water, but with sewage, the action is complicated.

As the reader may find on reference to a former chapter, and as, indeed, his own common sense must tell him, sewage holds in suspension finely-divided filaments of paper, cotton, linen, etc. These are gradually deposited in the interstices or tiny channels between the granules of the soil, and fill them up with what may be called a kind of *papier maché*. In fact, the upper layer of such a soil is water-proofed ; the sewage in consequence penetrates into it less and less readily, and begins to stagnate upon the surface in small pools.

Sewage, further, contains no small amount of fatty matter derived from soap. This also is deposited upon the soil and helps to render it less easily pervious to water. I have heard, on excellent authority, of an instance of a plot of good, dry meadow-land, perfectly free from any tendency to swampiness, which became part and parcel of an irrigation farm. Everything was done according to the rules of art, to promote drainage, and to prevent stagnation. Nevertheless, within a few years the character of the soil was so changed that it became a haunt of snipe. Surely every naturalist, and every sportsman must see the meaning of this fact.

Industrial waste waters may become especially injurious to the land if they contain large proportions of carbo-hydrates, that is of sugary, starchy, an gummy matters; such as the drainage of sugar-works, starch-works, distilleries, breweries, and kindred establishments. Such waters ferment and render the soil "sour." Before irrigating with such liquids, the compounds in question should be eliminated or destroyed.

A sewage irrigation farm, therefore, if the soil is not to suffer deterioration, must be ample in size so that each field, after being dosed with sewage for a year, may be disconnected from the pipes, ploughed over, and treated in the ordinary manner of farming for at least the next year, or, perhaps, for the next two or more years, whilst the sewage is made to flow over other plots. By such alternate treatment, the land may be kept from swampiness. But it must be admitted that such a farm would, from its size, be very costly, both to acquire and to lay out, and would compel the urban authorities to become farmers on a very large scale.

I come now to the question, In how far does sewage irrigation turn to good account the manurial matters; in other words, the plant-food contained in the sewage? The great objection to the old custom of running sewage at haphazard into the streams and rivers, as well as to Sir J. Bazalgette's modification, is the waste involved. The available and accessible supplies of phosphoric acid, potash, and combined nitrogen in the world are by no means infinite. If we go on year by year, and century by century, pouring them into the sea, a time must ultimately come when these essentials, if not exhausted, must become scarce. For lack of them the yield of the soil will decrease, and human food must in necessary consequence grow scanty. Hence one of the strongest recommendations of any system of sewage treatment would be that it utilises the whole, or practically the whole of the plant-food present in the sewage. This is the more important, since the very kinds of matter which are the most precious when applied to the soil, are the most dangerous and noisome if allowed to find their way into the water.

Now Dr. Maercker (*Zeitschrift fur Spiritus Industrie*, vi., p. 371) has recently been making some careful experiments on the waste waters of starch works. These waters are fully as offensive as town sewage, often, indeed, more so, and contain the very same manurial matters, viz., combined nitrogen, potash, and phosphoric acid. Hence their introduction into the rivers, without due previous purification, is in most countries prohibited by law. One of the ways in which their purification has been attempted is by irrigation, applied chiefly to grass-lands.

Dr. Maercker, it appears, found the total flow of waste water from a starch works, and by analysing average samples ascertained its composition. He then, in like manner, analysed the grass of the meadows and found its quantity. From these two sets of figures it appeared that of the total plant-food conveyed to the meadows only 2 per cent. of the potash, 3·2 per cent. of the nitrogen, and of the phosphoric acid only 2·9 per cent. were retained and utilized. The residual majority of these three valuable ingredients was therefore wasted!

This case may, indeed, be considered exceptional; but even Dr. Frankland, who is certainly not likely to undervalue the good effects of irrigation or to magnify its shortcomings, admits that of the total combined nitrogen distributed upon an irrigation field in the form of sewage, from one-third to one-half may make its escape in the effluent water, and be lost in the rivers.

But there is a further consideration: Is manurial matter, when dissolved in a large quantity of water, as is the case with sewage, presented to plants in a favourable form? I have already shown that irrigation, by lowering the temperature of the soil, delays the maturity of crops. But we may go further. Mr. E. Manson, C.E., in a work entitled *Sewage no Value; the Sewage Difficulty Exploded* (E. and F. R. Spon), makes the following admission: "Sewage, like water, retards the ripening of the fruit and grain, and develops the leaf. Sewage cannot supersede manure, for it cakes the ground, seals up its pores, and prevents the air from getting at the roots of the crops. It has been found at all times, and in all climates, that irrigation develops the growth of the leaves at the expense of the fruit or grain." These statements, which are fully in harmony with my own experience, are of the greater weight as coming from one who apparently favours sewage irrigation, and is no friend of chemical treatment.

Entering now upon the sanitary phase of the subject, we have to ask, Does irrigation render sewage, or water contaminated with sewage, sufficiently pure to be introduced without danger into any ordinary river? Does it, in effecting purification, give rise to any nuisance or injury to persons living near the irrigation plots? Are the vegetables grown on such irrigation plots fit for human consumption and for the food of animals whose flesh or milk may form a part of our diet?

To the first question an answer cannot be given in a couple of words. It may be at once granted that, providing always that an irrigation field is properly laid out, that the soil and subsoil are suitable, and have not become choked by too prolonged and continuous use, and that the sewage is not swept away into the drains and the streams by floods, the effluent water will, on analysis, show but a very small proportion of organic putrescible matter. In colour, transparency, and odour, it will probably not remind the spectator at all of sewage.

But it is now contended, on very good grounds, that the chief danger of the introduction of sewage into streams, wells, etc., lies not in *dead organic* matter, but in *living organised* matter. In other words, what we have to keep out of our drinking waters is not so much the decomposing residues of plants and animals and the excretions and secretions of the latter, as those minute living beings known scientifically as *bacteria, bacilli, micrococci*, etc., and called popularly "disease germs," from the fact that certain of them at least, when introduced into the body of a living animal, have the power of setting up morbid changes which may prove fatal.

We have, therefore, to ask whether irrigation or any kind of filtration can be depended upon to remove these disease germs? The answer must be in the negative. Dr. Percy Frankland, though a friend by right of heredity of sewage irrigation, made the following honest admission in a paper read before the Society of Arts on March 13th, 1884 : " There is absolutely no evidence that morbific matter, if present, would be removed," either by irrigation, or by " downward intermittent filtration," of which in a subsequent chapter. "On the contrary," he proceeded, "there is very strong reason to believe that these processes of purification offer no sort of guarantee that noxious organised matters present in the wage may not pass through into the effluent. For the removal of organic matter by means either of irrigation or intermittent filtration depends upon the oxidising action which a porous soil exerts upon such matter, and it is quite analogous to the purification of water percolating through a few feet of soil into shallow wells. Now, the instances on record of the percolation of sewage into shallow wells becoming the means of infection are so numerous and so well authenticated, that it is unnecessary for me to refer to them here. . . . . . At Stuttgart, in Germany, and Winterthur, in Switzerland, some years ago, epidemics of typhoid fever were proved most conclusively to have been caused by the contamination of the water supply with the effluent from irrigation meadows."

At the same meeting, Dr. Jabez Hogg, F.R.M.S., the distinguished microscopist, cited the well-known case of a stream in Switzerland which soaked through an entire mountain of oolitic rocks, yet, on emerging into another valley, was found to have brought with it the infection of typhoid fever.

I have seen flocks of sewage fungus (*Beggiatoa alba*) in the effluent from a large, well-managed irrigation farm on a deep soil. The spores of this fungus must, therefore, have traversed the soil.

In passing, I may here mention that this fungus is not, as many persons suppose, a certain indication of the presence of organic pollution. It flourishes where and only where sulphur is present. It is found in streams, into which fall sulphuretted mineral waters, as in the Pyrenees; it has been observed in the drainage from heaps of "tank-waste," the residue of the Leblanc alkali process, as well as in sewage, and sewage-polluted waters. Nor is its presence in sewage at all universal or apparently connected with the degree of pollution. It is sometimes abundant in the sewage of a town in one season, but in the next it may be almost or altogether absent, though no change can be traced in the character of the sewage. Instances of this have been noted in the sewage of Aylesbury as it arrives at the Sewage Works. Hence the presence, the quantity, or the absence of sewage fungus cannot, as it is popularly supposed, be taken as an index of the greater or less degree of impurity of the water.

Returning from this digression, we must conclude that if the living microscopic organisms in sewage are its most dangerous feature, and if irrigation is unable to remove such organisms, then the effluent from irrigation fields cannot safely be allowed to escape into wells or streams which supply water for human consumption. It seems not improbable that the chief mischief of excrementitious matters, etc., in water, is that they give scope for the multiplication of the "disease germs" in question.

It has next to be asked : Does irrigation effect its object without occasioning annoyance or injury to the inhabitants of the district?

I have never happened to visit or to pass near an irrigation field in warm, still weather without detecting an unpleasant smell. At Gennevilliers, near Paris, the odour, on calm, autumnal evenings may, without exaggeration, be described as abominable. The Prussian Commissioner, Lefeldt, when visiting Romford, found the smells emitted to be "mephitic in the most fearful sense of the word." Now, if I am right in my conception of what constitutes a nuisance, a loathsome odour, even though no definite disease can be traced in those who inhale it, is something against which the surrounding population has the clearest right to protest.

As regards the production of actual disease in the neighbourhood, diarrhœa, dysentry, typhoid fever, etc., the evidence is somewhat conflicting. In England, enthusiastic irrigationists maintain that—*e.g.*, at Croydon—no increase of disease or of mortality in districts bordering on an irrigation farm has been observed. But against such negative instances (which somewhat remind us of the thief who brought witnesses to swear that they had never seen him steal a horse) there is to be placed direct, positive evidence.

It is well-known that in India irrigation with ordinary river-water is needful in dry seasons; it is therefore practised on a very large scale. But, according to Markham, it has been found that the health of the irrigated districts is deteriorated, so that a committee appointed to re-consider the question proposed that a double belt of trees should be interposed between the irrigated fields and any adjacent villages.

Now, if irrigation is thus recognized as anti-sanitary in India, where ordinary river-water is employed, and where the fields are flooded or moistened only in time of drought, may we not expect at least as great injury from sewage irrigation in England, where the water applied is seriously polluted, containing sometimes disease germs, and where the fields are to be drenched from January to December?

It has been all along contended by sanitary authorities that a river polluted with sewage is productive of ill-health, not merely to persons who drink of it, but to those who inhale air which has swept over it. Be it so: I then ask why a plot of sewage-irrigated land, exposing as great as, or perhaps a greater surface than does the river in any one part, and giving evil odours, distinctly recognisable, should not be equally noisome? However rapidly the sewage may sink into the earth, a certain portion must escape by evaporation, and must rise into the air before it has had the opportunity to be disinfected.

To meet this difficulty, a clever hypothesis has been devised, the only flaw in which is that it lacks foundation. We are told that so long as a sheet of water or a stream, however full of disease germs, remains quiet, these germs will not rise up into the atmosphere. But if, as is usually the case in foul ponds or streams, fermentation is going on and bubbles of gas are rising to the surface and bursting, these microscopic organisms are flung upwards into the air and are carried away by the wind. But from the surface of a plot of land, wet with sewage, such bubbles do not arise, and consequently the disease germs present do not escape into the air!

In reply to this supposition, I must remind the reader that if we pour any liquid upon a portion of earth somewhat dry—for instance, the soil in a flower-pot—we see bubbles arise and force their way through the film of liquid, until all the air existing in the interstices or pores of the soil is completely expelled. These bubbles must, of course, on bursting, project any germs present into the air, just as do the bubbles bursting on the surface of a polluted river.

But the assertion that germs are not thrown off from the surface of still water without the action of bubbles is denied on the faith of direct experiment.

Portions of liquid containing organisms of known kinds, and not undergoing any fermentation, have been placed in dishes under glass bells. On standing overnight, organisms of the same kind as those present in the liquids have been found in plenty upon glass plates suspended over the dishes, but under the bells. Hence, in accordance with our present state of knowledge, the action of a polluted river, and of a plot of land, frequently or constantly moistened with sewage, must be very similar—substantially alike.

I must now invite the reader to make a very easy experiment. If he has a garden, let him select a plot of ground, say a square yard, free from vegetation, and, if possible, of a fairly open texture. Let him then pour over this plot a quart of sewage, or, in default of that liquid, a similar quantity of the mixture of urine and soap-water from the chambermaid's sloppail. Let him then watch the result. If the weather is at all genial a number of two-winged flies (*diptera*) of different species will soon settle upon the wet earth, and may be seen sucking up the moisture. As the first comers fly away they will be succeeded by others, and for some hours the damp plot will be a source of attraction to these unclean insects. A similar plot of ground, moistened with clear water by way of a counter-check, will, on the contrary, have very few visitants. This experiment proves that, its admitted deodorising and disinfecting powers notwithstanding, earth does not immediately absorb and destroy the offensive matters of sewage. Secondly, we learn that soil thus moistened, and, *a fortiori*, any sewage irrigation field, is an attraction and an encouragement to flies. Let us take the case of a farm receiving the sewage of a town. All the summer long it will be haunted by numbers of blow-flies, dung-flies, house-flies, gnats, blood-suckers (*Stomoxys calcitrans*), and many others. All these will become saturated with the putrescent matter. If the sewage contains, as it conceivably may, the excretions of a cholera or a typhoid patient, the flies imbibe the "germs" of such diseases.

Some of the insects will then enter our houses, and crawl over articles of food. Others settle upon our persons and inflict malignant wounds. Fatal illness has not unfrequently been traced to the bite of flies which feed on sewage or carrion. These flies being now recognised as among the greatest agents for carrying putrid poisons and disease germs to the healthy, it is important that all places where they can increase and multiply, and all matters upon which they may feed, should be made offensive to them or destroyed, as the case may admit. Thus, we bury the dead bodies of birds, beasts, and fishes, which are either uneatable or which have perished from disease, as well as all putrid and putrescible solid matter. We do not, if we are prudent, allow blood—especially the blood of animals slaughtered by way of stamping out the cattle-plague—to stagnate or to soak into the earth. To collections of excrementitious matters, such as the contents of cess-pools, we add such disinfectants as may make them unfit for the habitation of maggots. We do not throw the dejecta of cholera patients, etc., into the gardens and the fields, but treat them with corrosive sublimate. Thus, in all cases where we have to deal with offensive animal matter, especially such as has been the seat of virulent disease, we withdraw it from the reach of dipterous insects by fire, by chemical agents, or by burial. But if we practice sewage irrigation we spread out excrementitious matter, possibly containing disease germs, over a large extent of surface, and thus throw it fully open to unclean insects.

Lest anyone should think that the views above given are exaggerated, we may refer to a recent communication by Dr. Maddox, in a late number of the *Journal of the Royal Microscopical Society.* This gentleman has proved that the "comma bacillus," now known to be the agent producing cholera, "can pass in a living state through the digestive organs of flies, which may in this manner become carriers of contagion." Dr. Grassi, of Rovellarca, studied this question two years ago. His experiments point to the conclusion that flies may be regarded as veritable promoters of epidemics, and agents in the propagation of other infectious maladies. To prove this, he placed upon a plate in his laboratory some ova of the *Trichocephalus*, and in a short time these were removed by the flies, and deposited in another place some little distance away. He caught some of the flies, and found that their digestive tubes were full of feculent matter, and of the ova. He also put segments of the tape-worm (*Taenia solium*) into water. Some of their eggs remained suspended in the water; the flies drank of the fluid, and in less than an hour he found the eggs of the tape-worm in their intestines, and in their excrements. Flies, he further found, will also transmit the eggs of the small threadworm *Oxyuris*.

He likewise fed flies on mildewed cream, and afterwards found within them, *Oidium lactis*. Other flies fed on the "muscardine" of the silkworm voided spores of that pest in their excretions still capable of development.

Hence Grassi inferred that these organisms are not destroyed when swallowed by insects, since the germs of mildews and of *Schizomycetes* pass through their bodies uninjured. Flies are also certain to carry about with them living organisms on their feet and trunk.

M. Daraine shows that by feeding flies with infected blood they can be made to convey infection.

Dr. Manson has likewise shown that mosquitoes are carriers of the germs of *Filaria sanguinis hominis*, and that it is quite possible for the dreaded "Tsetse fly" of Central Africa to transmit infections to the animals which it attacks.

After many and varied experiments, Dr. Maddox found that the cholera-bacillus can pass through the intestines of a fly in a living state.

Hence, surely anything which encourages flies and supplies them with putrid and frequently diseased nutriment ought to be carefully avoided.

We now come to the third and last question concerning the sanitary value of sewage irrigation : Are the vegetables grown on sewage farms fit for human consumption or for the food of cattle?

On this subject very conflicting opinions prevail. I must remark that the question probably turns on a point commonly left quite out of consideration. We know that from time immemorial night-soil, the material from cess-pools, including all the constituents to be found in the sewage of a residential town, has been applied to gardens without the production of any known evil effects. At least, if mischief has arisen it has not been recognised. But such night-soil, as far as I am aware, was either dug into the earth some time before the intended crop was planted or else was trenched in between the rows, say, of potatoes, apple-trees, gooseberry-bushes, etc. It was never applied in a fresh state to strawberries, celery, salad-herbs, and the like. Thus the fæcal matter never came in immediate contact with the roots of plants, nor was the application repeated in the same season. In farming, night-soil was, and doubtless still is, applied to stubble-fields, to be ploughed in, and scattered over pastures and meadows in the late autumn. But I have observed that cattle turned into such pastures next spring would avoid those parts where the night-soil had been applied, and keep, in preference, to any part which had been dressed with farm-yard manure. Now, as regards sewage irrigation all the circumstances of the case are altered. Excrementitious matters suspended or dissolved in water are passed into the fields not merely during the absence of crops or between the rows of plants, but, practically speaking, throughout the season and in all parts of the land. Suspended excrement, healthy or diseased, as the case may be, comes in direct and constantly renewed contact with the roots and the stems of grasses and other plants, and no matter how close upon perfection the drainage of the field may have been carried, there it will adhere. Of this any candid person may convince himself by observation in such fields, or even by direct experiment. Let him, for instance, fit up a funnel loosely with a grass root and pour sewage upon it. However well the filter acts, he will soon find the stems and the roots coated with an adhesive matter, concerning the nature of which no doubt can be entertained.

If, then, such grass is cut from time to time, without

any intermission of irrigation, and is given to cows, the fæcal matters in question are conveyed into the stomachs of these animals, and their tissues and their secretions *may* become poisoned. That disease germs may pass into milk and thus reach fresh victims is an established fact.

But we have to do not merely with sewage matters clinging to the outer surfaces of plants. The impurities penetrate also into the interior. Herr Lefeldt, to whom I have already referred, in his report on the various systems of sewage treatment as pursued in this country, notices stems of grass from irrigation meadows, full of unassimilated sewage matters (*Kloaken-Stoffe*). Were the irrigation suspended for a sufficient time these matters would doubtless be assimilated by the plants, which would then be perfectly harmless. But if the sewage flows on day by day, fresh excrementitious matter is absorbed as fast as—perhaps faster than—the former doses can be assimilated. Surely such grasses or other plants placed in similar conditions must be of very doubtful value as food, whether for man or beast.

I now come to the experiments carried out by Mr. Smee, Jun., and published by him in a work entitled, " Milk in Health and Disease." These experiments and their results have been met, I am compelled to say, with something very like the "conspiracy of silence." Certainly, so far as I can learn, no attempt has been made to prove them inaccurate, or to refute the author's conclusions. Two cows were set aside for experiment. The one, which we may call A, was fed on sewage irrigation grass, and the other, B, on grass from an ordinary meadow. The milk obtained from each cow was kept separate, and examined. It was found that the milk of A became not merely sour, but it putrefied and stank much sooner than that of B. It was noticed that a favourite cat, exceedingly dainty in its tastes, entirely refused to lap the milk of A. The butter from A's milk became rapidly rancid as compared with butter obtained from cows fed on ordinary pasturage. Cream from the milk of A required, in three successive lots, $1\frac{1}{2}$ hour, $1\frac{1}{2}$ hour, and $2\frac{1}{2}$ hours to churn, and the butter was soft and smeary. Check samples of cream from cows fed on normal food required only thirty-five minutes, $1\frac{1}{2}$ hour, and $\frac{3}{4}$ hour to churn, and the butter was firm.

So far, of course, this experiment is open to the objection that the bad quality of the milk and butter from A was due to some morbid condition in herself, rather than in her food. To meet this doubt, Mr. Smee reversed the experiment, feeding B on sewage grass and A on normal herbage. He also tried other cows. Still the results reached were practically the same, the milk from every cow fed on sewage grass was notably more prone to putrescence than that from cows fed on common meadow grass.

Mr. Smee made further experiments on the grass itself. He found that the juice of sewage grass became more quickly and more offensively putrid than that of common grass. Hay made from sewage-grass, if kept in a vessel of water in a warm place, quickly set up a putrid fermentation, whilst hay from ordinary grass, treated in the same manner, behaved quite differently.

It will, of course, be granted that the application of sewage to fruits and vegetables which are ordinarily eaten raw, such as celery, lettuce, watercress, radishes, strawberries, etc., requires the greatest degree of caution. In connection with this subject it should be remembered that when it not long ago seemed probable that the cholera might visit Paris, the inhabitants were formally warned by the sanitary authorities against consuming vegetables from the sewage irrigation farm at Gennevilliers!

I should suggest that where sewage irrigation is practiced, it should be suspended for the very least a fortnight before the crop is reaped or gathered. Thus, in case of strawberries, it might be well to shut off the sewage from the appearance of the blossoms until the plants have ceased fruiting.

I cannot help expressing my regret that the warm friends of irrigation should have shown so little disposition to investigate this part of the question more closely. Have they a secret misgiving that the truth, if ascertained, may be of an unpleasant character?

In conclusion, I would submit that irrigation, though an excellent method of disposing of, and at the same time utilising, sewage, where suitable land is available, where the climate is warm, and the rainfall scanty or intermittent, is not applicable where these conditions are absent. Any attempt to represent it as the only means of dealing with the sewage difficulty, and to force it upon reluctant communities, is a grave error—in fact a crime, the motives for which are in most cases hard to trace.

CHAPTER VII.

MODIFICATIONS OF IRRIGATION.

IN the last chapter I have understood irrigation as consisting in the direct application to the land of sewage as it flows from the town without any previous preparation. There are, however, modified processes, some of which obviate more or less completely certain of the inconveniences attending upon irrigation "pure and simple."

One of these is the introduction of settling-pits, into which the sewage first flows and collects before being passed upon the land. These pits serve, in the first place, to average the sewage. It has been already shown that the sewage of any town varies considerably both in strength and general characters at different parts of the day and night. By averaging the whole, its treatment, on what system soever, is made more convenient.

But the settling-pits have a more important task to perform. I have already mentioned, in passing, as among the ingredients of sewage, grit and silt derived from the streets and roads, and I have pronounced this silt as a formidable difficulty in any and every method of sewage treatment. It is, of course, most abundant after heavy rain, but it is also much affected by the character of the streets. If these are asphalted, paved with wood, or even with stone, the quantity of such silt is relatively trifling. But where the abomination devised by Macadam prevails, as in all parts of London except the City, its quantity is something frightful.

Sewage, if allowed to stand even for a few hours in a settling-pit deposits the greater part of this grit. But unfortunately, each grain, by sojourning in the sewer in company with noisome matter, acquires an organic coating of most offensive odour. Hence such settlings cannot be used in mending roads, in making mortar, or in filling up hollows in the ground. The only legitimate use to which they can be put is the improvement of heavy clay soils, and if such are not close at hand their disposal is not easy. On the other hand, if sewage is allowed to flow upon the land without having first deposited its silt and grit, the channels are being continually choked up, and the entire level of the ground will be gradually raised. In filtration and precipitation processes the silt is equally an inconvenience.

Hence there is a strong temptation to use settling-pits, even when the disposal of their contents is difficult. Their odour is not pleasant, and has, indeed, been described by a foreign authority as being "mephitic, in the most fearful meaning of the term." Still, this difficulty may be got over by the use of some suitable disinfectant.

Another modification of the sewage irrigation process is the introduction of some such disinfectant prior to its being allowed to flow over or through the land. This plan has been adopted at Carlisle, the material used being a combination of carbolic and sulphurous acids. By this expedient certain of the objections to crude irrigation are got rid of. Thus the disagreeable smell often given off by irrigated land is done away with; the plague of flies is greatly reduced; and cattle eat the grass with a better relish. Whether the dairy produce obtained from cows fed on such disinfected sewage grass is equal to that of cows pastured on ordinary grass has not, so far as I can learn, been experimentally tested.

It is remarkable that attempts in the disinfection, or at least de-odorizing, of sewage prior to use on land have not been more numerous and more varied. It must be remembered that the choice of disinfectants for this purpose is limited. Some are too costly, some poisonous. The use of carbolic acid is not quite free from objections, as fish in rivers into which it is introduced seem to be rendered more liable to parasitic diseases. Experiments on this subject are needed. I should gladly have undertaken this inquiry, but I have been advised that it would be legally unsafe in England without a licence under the "Vivisection Act"!

Irrigation has been occasionally tried as a supplement to some of the precipitation processes to be discussed in a succeeding chapter. Such a combined method is, or was recently, in operation at Coventry. The notion which underlies such schemes is that of a smaller outlay for chemicals and plant than would be required for a thorough precipitation process; and, on the other hand, that a smaller plot of land would suffice to receive the sewage. Whether there is any real economy in this double working is, however, at least open to question. Just in proportion as a sewage is purified by precipitation, its manurial value for irrigation will decrease down to that of plain water. A certain chemist, now no longer living, did, indeed, once try to argue that the effluent from a precipitation process which he had pronounced well purified was yet better suited and more valuable for irrigation than the original sewage. This attempt at running with the hare and holding with the hounds met with little approval.

Sewage freed by some precipitation process from its suspended impurities will, however, have the advantage of not choking up the pores of the land, and thus rendering it liable to become water-logged and ultimately swampy.

CHAPTER VIII.

FILTRATION.

FILTRATION differs in principle but little from irrigation. During the winter an irrigation field is, in fact, merely an extensive earth-filter. In summer, the difference is greater, as in the irrigation field the growing plants take a certain share in the work, and absorb a part of the impurities existing in the sewage.

The duties of a filter are firstly, to remove mechanically all matters, organic or inorganic, suspended in the sewage. This work is often known as "clarification." Secondly, the water, in descending into the pores of the filter bed, draws along with it a certain quantity of atmospheric air, which oxidises, or, in plain language, burns up the animal or vegetable matter present, converting it into substances which are no longer injurious to health. Thirdly, the process of filtration decomposes a number of organic compounds quite irrespective of any oxidation. Lastly, it is supposed, or at least hoped, to withdraw from the liquid to be filtered any disease germs which may be present. These tasks are performed in very varying degrees, according to the kind of material used, to the time the filter has been kept in action, and to the rate at which it is intended to work.

Whatever be the material chosen, there are two different kinds of filtration: downwards and upwards; or as they are also named, descending and ascending. In downward filtration, which is by far the most common, the liquid to be purified flows upon the surface of the filter bed, passes down through its mass, and is delivered at the bottom. In upward filtration, which is practicable only under peculiar circumstances, the liquid is forced up through the filter bed, and flows out at the top. Downward filtration is not only the simplest and oldest kind of filtration, but it is generally the most efficient. In it only is the liquid purified in the second method abovementioned, that is, by means of atmospheric air drawn down along with it into the pores of the filter. But to this end it is necessary that there should not be any considerable depth of water standing above the level of the filter bed, as otherwise the aëration cannot fully take place. The upper surface should be kept only just wet, a layer of a quarter inch in depth being quite enough.

The manner in which the filter is made up is very important; if the bed is too thin and irregularly thrown together, the water is apt to force its way through in some few places, and thus to escape unfiltered. In building up the filter, the coarsest materials are spread out at the bottom; over these are placed, regularly and evenly, layers of finer and finer materials, the finest being near the top. But lest the inrush of the water should disturb this fine upper layer, a few large, flattish stones, pieces of slate, or large lumps of coke are laid on the top of all in order to receive the stream at its entrance, and distribute it in thin layers over the surface. It is further desirable that the liquid to be filtered should enter in a shallow sheet over a wide lip. Roughly speaking, for a filter of given size and material, the less rapidly the liquid passes through it, the more thoroughly will· it be purified.

The chief materials used for forming filter beds are gravel, sand, moor-earth, burnt clay, pumice, coke, animal charcoal, wood charcoal, peat (raw or carbonized), seaweed charcoal, spent oil shales, Kimmeridge carbon as obtained from the so-called Kimmeridge blackstone, lignite, spent dye woods and sawdust, especially if slightly carbonised by fire or by the application of acids; shavings, rushes, faggots, straw, and other hard vegetable matters exposing considerable surface. Further, I must mention spongy iron, magnetic iron ore, black oxide of manganese, scrap iron, and cylinders of unglazed porcelain. These materials are used sometimes alone, sometimes in combinations, which may be varied almost indefinitely.

It will be easily seen that for the treatment of sewage many of these materials are out of the question. As it has been already pointed out, the filtration of sewage is far more difficult than that of spring, river, or lake waters for domestic or manufacturing purposes. I lately found that half a litre (about $17\frac{1}{2}$ fluid ounces) of the effluent from a certain sewage works took exactly two hours in passing through a 5-inch paper filter. The same measure of the untreated sewage from the same works took exactly seventy-two hours to pass through a filter of the same size and quality. This simple experiment shows the relative difficulty of filtering sewage. The large quantity of paper pulp and of the fibres from textile goods always present in sewage, even such as appears almost transparent, quickly clogs up the filter beds. The fatty matter derived from soap and from culinary operations combines with these fibrous matters in—as it might be called—waterproofing the filter, forming an impervious, water-repelling layer not only over the surface, but even within the interstices of the mass. This choking and clogging process is quite independent of the nature of the materials used, or rather, we may say, that the better the filter was originally, the more thoroughly and quickly it will become choked; whilst one which allowed the water to run through rapidly will remain in full activity much longer. Hence filters for sewage must be relatively larger than those for river water, etc., and they will require more frequently cleaning or changing. Cleaning a filter bed when it has become dirty is in many cases a most troublesome operation, as the fæcal matters which adhere to the sand, coke, etc., will have become very offensively putrid. It is commonly said: take out the soiled and clogged materials, and spread them out to the air. But for this the necessary room is not always to be had, and the smell of the mass thus exposed is liable to raise objections. If we wash the soiled matter, where are we to turn the washings? If into the river, we occasion a pollution nearly as great as if we had all along allowed the sewage to flow in unfiltered; if we turn it back into the sewage, we have the work to do over again under greater difficulties than at first. Purification by fire is with some materials impossible, and in most cases it requires special kilns or furnaces.

A change of filters, to be used alternately, is recommended where the room is sufficient. When one filter shows signs of clogging, or allows the sewage to pass through in a foul condition, it is shut off, and the liquid is turned into the other.

Sand and gravel, however deep in mass, and however well spread, do little more than remove the suspended, or, I might say, the coarser suspended impurities. But they are relatively cheap, procurable almost everywhere, and when foul they may be easily dealt with. The sand may serve to improve clay soils, whilst gravel, after due airing, can be used for mending field paths; but *never for filling up hollows upon which houses may subsequently be built.*

Moor-earth—a mixture of sand and peat—where procurable, acts better than sand or gravel. When spent, it may be used for farming and gardening purposes. Burnt clay ranks about with gravel. Pumice is tolerably efficient, but it is too expensive. Animal charcoal, otherwise known as bone-black and as spodium, is also far too costly. This is the less to be regretted since, whilst it purifies polluted waters well at first, it afterwards becomes charged with offensive matter in such a manner that it contaminates water instead of improving. Wood charcoal, if thoroughly burnt, so as to be quite free from oily and tarry matters, acts fairly well in combination with sand, etc. Peat was applied to the filtration of town sewage by the Peat Engineering Company, Limited. It acts well for a time, but when clogged it is difficult to deal with. Peat-charcoal is a promising material, as seaweed charcoal would be also, could it be procured at a lower figure. Mr. S. K. Page, manager of the Aylesbury Sewage Works, has made a series of careful experiments upon the Kimmeridge carbon. Here the results were at first all that could be desired, but the bed soon choked up, and the first cost of the material would not permit it to be frequently exchanged. Unglazed porcelain filters, as proposed by Pasteur and Chambeland, may be considered perfect, as they remove even microscopic organisms. But for dealing with the sewage of a town they are utterly out of the question.

Spongy iron bears deservedly a very high character,

and is even said by some authorities to remove bacteria, though this is denied by Mr. Jabez Hogg, M.R.C.S., F.R.M.S., one of our most competent microscopists, and, in this special department, perhaps the most eminent. Be this as it may, spongy iron, like all metallic substances, is inapplicable to the sewage of manufacturing towns, as it would be acted upon and injured by the acids and acid salts rarely absent. It would also, like manganese, soon become mixed with the silt and grit brought down by the current. To separate this refuse matter when the filter-beds require cleaning would be no easy matter without wasting a considerable proportion of the active material.

There are two methods, or processes, of filtration which have been especially recommended for dealing with sewage. One is that of Professor Henry Robinson, first made public at the Dublin meeting of the Sanitary Institute, and reported in the *Sanitary Record* for October, 1884. This gentleman—an engineer, if I mistake not—proposes to adapt clay lands for a something midway between filtration and irrigation by digging out the soil to the depth of six feet, burning it, and arranging it in layers interspersed with a stratum of open, alluvial soil, of course unburnt. Such a bed, six feet in depth, will, we are told, continuously *clarify*—it is not said *purify*—the sewage of 1,500 persons per acre. The cost of preparation however, reaches, according to Mr. Robinson's own estimate, the modest sum of £750 to £1,000 per acre. Every one knows that such estimates fall far short of the actual cost of these schemes when reduced to practice. An observant friend tells me that every £1,000 in an estimate means in reality about £2,500. But let us accept the estimate, and take a town with a population of 300,000 persons—say, Leeds. At 1,500 persons per acre we should require 200 acres of land, to be prepared at a first cost of £200,000! Or, for London, 2,666 acres, costing £2,666,000. I do not find whether in this estimate is included the purchase of the land, which anywhere within a moderate distance from a large town must be a heavy item. But this is not all; Professor Robinson is of opinion that before running the sewage upon this bed the coarser suspended impurities should first be removed by a process of ascending filtration, of course through a separate set of beds, the materials of which, when sufficiently polluted, are to be taken away, and "dug into low-lying land," whilst fresh ascending filters must be brought into action. These ascending filters and the land for digging in the coarser sewage matters will occupy a considerable space! Even if we are bold enough to assume that the acres stated will go on for ever purifying the flow of sewage, we must admit that the yearly working of this most singular scheme will be no trifle. Nor must it be forgotten that the plots of land thus prepared will be thereby converted into deserts, which neither art nor nature will henceforth be able to reclaim.

We come now to the process, designated officially as "intermittent downward filtration," and recommended officially and officiously as *the* alternative to irrigation, to be used when land of a quality and in a situation suitable for a sewage farm is not to be had. That the term in question is novel must be admitted, but that there is any novelty in the thing itself it might be hard to show. The reader has been already reminded that all ordinary filtration is "downward," and the term intermittent merely implies that the filter beds are to be made in duplicate or in triplicate, the one to be in use whilst the others are being cleansed.

It may be well to state Professor Frankland's original plan in full. A plot of ground is to be procured, say three times larger than is necessary to receive the entire sewage of the town. The soil of this plot, to begin with, must be of suitable quality, and it is prepared (presumably dug up, drained, and rendered uniformly light and open), to the depth of six feet. It is then divided into three plots. Upon one of these the entire sewage is run for eight hours. It is then turned upon the second plot for the next eight hours, and finally upon the third plot for the last eight hours. Thus each plot will have eight hours' action and sixteen hours' rest. I am not aware that there is any sacredness in these exact numbers. It would, perhaps, be permissible to divide the plot of land into four parts instead of three, thus giving each part only six hours' action and eighteen hours' rest. From certain laboratory experiments the distinguished author of this scheme calculated that each acre of land would in this manner purify the sewage of 3,300 persons *in saecula saeculorum*. In a recent letter to the *Times* he has reduced his estimate to 2,000 persons per acre, and some of his disciples have further varied the figure. Another point of difference is regarding the possible utilisation of the land laid out as filtration beds. Professor Frankland, unless I misunderstand him, holds that it involves the sacrifice of the land from an agricultural point of view, and, of course, of the fertilising matters contained in the sewage. Mr. Bailey Denton—who seems to be to Professor Frankland what Ali was to Mahomed and Mr.

Grant Allen to Charles Darwin—thinks that this peculiar kind of filtration is the best means of getting a good crop. Now I would ask any practical farmer what crop—save, perchance, rye-grass—would be the better for having turned upon it, in addition to the natural rainfall, the sewage of 3,300 persons; that is, on the average 99,000 gallons of water per acre every day from year end to year end?

Professor Frankland has now essentially modified his views, and recommends that the earth for a "downward intermittent filtration" bed should be prepared to the depth of *two* feet only, thus tacitly admitting that the lower four feet are of little use, and that his former experiments, or at least the conclusions based upon them, were fallacious. It would be a grave error to cavil at a man of science for retracting and modifying opinions which, upon further experiments and observations, are found no longer tenable. But surely a *savant* who thus openly and honourably confesses his own fallibility might be led to inquire whether some of his other utterances are not quite as much in need of revision?

We have now to ask whether a daily rest of sixteen, or even of eighteen hours will keep the land up to its original degree of permeability? We have already seen how the soil is waterproofed by the deposit on its surface and in its pores of the fibres of paper, linen, cotton, etc. Now these fibres, if exposed to air and moisture, would doubtless in time become disintegrated and oxidised, save for one circumstance. That is, they are coated with greasy matter, as it has been mentioned above, and are thus protected in a very great measure from the action of water and of atmospheric oxygen. Even ordinary clean water applied to soil in such proportions as it is recommended for "intermittent downward filtration" will ruin its texture. Let any one take a box of soil of one foot square, set in it any kind of plants other than swamp vegetation, and then pour upon it, day by day, two gallons of water (= 90,000 gallons per acre), he will find that its porosity will be destroyed, and that the plants will not thrive. How much more must this mischief occur when we have superadded to the water the waterproofing materials above mentioned, and when we have, in addition, the yearly rainfall to contend with? In proportion as the soil becomes waterlogged (which must ultimately be the case), little pools will begin to stagnate upon the surface, and a larger and larger proportion of the moisture will have to be got rid of by evaporation—a result injurious both agriculturally, as chilling the soil, and from a sanitary point of view, as favouring the diffusion of sewage vapours and probably of disease germs. "Digging in" the deposit cannot well be executed whilst any crop is on the ground, and can at most only defer the evil.

It must never be forgotten that, whilst irrigation and precipitation present at least the possibility of some return, all the outlay in filtration, as in the Bazalgette system, is pure waste. The object is not utilization but destruction.

It must not, however, be supposed that filtration, even as applied to crude sewage, is in all cases to be condemned. Let us suppose a town or village which is not closetted, and where, consequently, little excrementitious matter finds its way into the sewers. But let there be instead some large manufacturing establishment, emitting much liquid refuse, acids, solutions of metals, drainage from tank-waste, residual liquors from dye-works, tanneries, paper-mills, etc. If we irrigate with such liquids we kill, instead of nourishing, the crops. If we precipitate we may obtain a clear, colourless, inodorous effluent, but the precipitate will be fit "neither for the land nor for the dunghill." In such cases filtration may prove the least objectionable method of treatment.

Further, filtration may often be usefully applied as finale to a precipitation process where extreme purity of effluent is desired.

CHAPTER IX.

PRECIPITATION.

THE processes for the chemical treatment of sewage are so numerous, so different in their principles, and so varied in their grades of efficiency, that a complete discussion of each is purely impossible. Fortunately, many of them are merely changes rung upon comparatively few agencies, so that they may be dealt with in groups.

Before considering the substances used to effect precipitation, we must prepare to meet a cavil which has been uttered with great confidence, not to say rashness, though its first author, like many inventors, is unknown, blushing, perhaps, at the fame which might be his mede. It is said that " chemical agents, though they may *clarify* —*i.e.*, withdraw suspended impurities—cannot *purify*, —*i.e.*, they are unable to remove dissolved impurities." It is difficult to understand how any chemist can make such an assertion without placing himself in a very unpleasant dilemma. To begin, it is well known that suspension and solution fade away into each other by scarcely perceptible gradations.

It is further evident that suspended organic matters —under which head we must include living organisms, the germs of disease, and putrefaction—must, in the strictest sense of the word, rank as impurities. If we remove them we *pro tanto* purify the water.

It is also known that there are two methods, if not more, in which dissolved substances can be separated from the solvent liquid: precipitation in the strict sense of the word, and occlusion or aborption. In a well-planned precipitation process these two modes of action are systematically combined. In precipitation proper the dissolved body enters into a definite and more or less stable chemical combination with some other substance introduced, forming a compound which is insoluble in the liquid, and which then subsides to the bottom. This occurs with organic as well as with inorganic bodies. Thus, if we dissolve a little white of egg (albumen) in water, and add a solution of sugar of lead, the albumen combines with the lead to form an insoluble mass. Or if we take a solution of gelatine in water and add to it a solution of tannin, the gelatine coagulates, and is deposited in combination with the tannin. Everyone acquainted with the arts of dyeing and tissue-printing will know instances where dissolved organic substances are rendered insoluble by contact with metallic salts.

These facts are here mentioned, not as in any way novel, but merely to prove the general proposition that dissolved organic compounds are capable of being precipitated. Before passing to instances proving that sewage matters in particular are capable of being thus thrown down, it is proper to refer to *occlusion.* Here a dissolved substance is withdrawn from solution, not by forming a definite chemical compound with some other body introduced, but by becoming entangled in its pores. Gelatinous silica and hydrated alumina, when freshly precipitated, are capable of thus entangling dissolved organic matters.

We may now pass to an experiment showing the precipitability of dissolved sewage matters. Take a pint of sewage and filter it through the finest filter-paper, which will prove a work of time. The liquid running through may be almost as bright and clear as spring-water. But if there be then added to this liquid a few drops of a solution of alum, or of hydrated aluminium chloride, commonly spoken of as muriate of alumina, or of sugar of lead (all which must be free from any excess of acid), there will soon form a white cloud in the liquid, which will then, before long, settle to the bottom. This is a compound of alumina (or respectively of lead) with the organic matter which was previously in a state of solution. If some of the filtered liquid is analysed before and after the addition of the alum or other precipitant, it will be found that a large proportion of the organic matter previously dissolved in the sewage has been removed. Thus, Professor Dewar, F.R.S., and Dr. Tidy, in the report of their recent prolonged investigation of the sewage of Aylesbury, and its treatment by the " A. B. C. process," state that they found about 60 per cent. of the dissolved organic matter in the sewage was removed, and, further, that the portion not thus removed was precisely that least likely to enter into offensive or dangerous decomposition.

Lastly, pure water is a thing which does not occur in nature, and which the majority of those who utter the above-quoted cavil have certainly never seen. If they mean to say that sewage cannot be brought to this state by chemical means—or, indeed, by any other—they merely utter a gratuitous truism. If under cover of this truism they insinuate that sewage cannot, by precipitation processes be so far purified as to be safely admissible into the rivers of a populous and cultivated country, they state at best an opinion which they would find it hard to prove.

It is, therefore, sincerely to be hoped that we may have heard the last of the cavil that organic impurities dissolved in water cannot be got rid of by chemical treatment.

I find, however, that when I wrote the above paragraph I was paying an undeserved compliment to public intelligence and candour. A few days ago there fell into my hands the prospectus of a new sewage process by aëration. The inventor goes somewhat out of his way to attack precipitation, and writes that the Royal Commission on the Metropolitan Sewage (1883) "admits"—note this term!—that the best precipitation processes only clarify, but do not purify! It is lamentable when a Royal Commission "admits" something very far from the truth. That Commission refused, or neglected—and in such a case, these two terms are nearly equivalent—to examine fully and fairly into the merits of precipitation. It overlooked the fact that many chemists, engineers, and other experts, who a few years ago were decidedly hostile to chemical treatment, have latterly seen reasons for changing their views. It would not, or at least did not, visit Aylesbury. It was satisfied to condemn precipitation on the faith of the archaic reports of the Royal Rivers Pollution Commission, reports which, if true at the date when written— and this is a fairly strong concession—are demonstrably false if applied, *e.g.*, to the process now in operation at Aylesbury.

It must now be asked, What organic matters are

capable of being eliminated by so-called chemical processes? We may say that this is the case with albumen and analogous compounds; with gelatine, mucus, and with peptones. The same holds good with pyin, the albuminoid constituent of pus. Thus we see that the most important portions—*i.e.*, those most likely to occasion mischief on decomposition—of blood, urine, and of the soluble part of solid excrements, are amenable to precipitation. That suspended matters can be precipitated has never, probably, been seriously disputed, though some of them, such as fatty particles, are much more difficult to deal with than the bulk of the dissolved impurities. Phosphoric acid is invariably present in sewage, being introduced by urine and by blood. Though not to be called an organic impurity, it is highly objectionable, as it is necessary to the growth and multiplication of disease germs and other microbia. It is satisfactory to know that it may be almost absolutely removed by precipitation.

In the waste liquors of industrial establishments there are a vast number of impurities capable of being removed by chemical treatment. Such are the salts of the heavy metals, colouring and tanning matters, etc.

Among the substances which resist precipitation are oils and fats, essences, waste products of gas-works, and refuse resulting from the manufacture of india-rubber articles; ammoniacal salts, salts of the alkalis, especially nitrates, nitrites, and common salt. Concerning these various bodies it is to be noticed that gas-works refuse cannot be lawfully run into the sewers at all, and that where soapsuds are abundant they are generally kept back by the manufacturers, and separately treated for the recovery of the fatty matter.

One of the principal animal products which hitherto has not been found practically precipitable is urea. This substance, in contact with a ferment which is never absent in urine, is quickly converted into ammonium carbonate, and never reaches the sewage tanks.

The ammoniacal salts and the alkaline nitrates and nitrites are never found either in polluted rivers, or in sewage in such a proportion as to be in themselves a nuisance. At the same time we must regret that they are not precipitable, because:

1. If they could be thus arrested, they would greatly increase the manurial value of the deposit obtained.

2. Though in themselves inodorous, incapable of putrescence, and, in fact, harmless, the nitrogen which they contain may be brought into putrescible conditions by the action of living organisms, and because they will doubtless favour the multiplication of microbia. It must be remembered that in irrigation and filtration nitrites and nitrates occur in abundance in the effluent water.

I have now to point out what are the properties, positive and negative, which a precipitating agent ought to have over and above mere efficiency. In the first place it must be cheap—cheap, not merely by reason of present small demand, but of abundant, or, rather, unlimited supply.

It must not be actively or cumulatively poisonous. Certain salts of lead are well adapted for throwing down organic matter from suspension and from solution. But, on the one hand, the deposit would be unsafe as a manure; and, on the other, the slightest excess remaining in the effluent water would be deadly to vegetation, to fish, and to cattle, which might drink of the stream. Antimony is excluded on the same grounds, as are also tin and bismuth, which are, in addition, too costly.

The precipitating agent must not have in itself any distinct colour, or generate a colour with substances which it may probably have to encounter.

This condition condemns, under most circumstances, compounds of iron. Many of these are in other respects well adapted for precipitation. But waters to which they have been added take a greenish-yellow colour on prolonged exposure to the air, and a yellow, ochreous deposit is formed on stones, brick-work, piles, etc. Though these deposits may be perfectly harmless, yet to the public they convey the notion of an excrementitious origin, and the process is at once condemned. Iron sediments, if containing sulphur, or if coming in contact with sulphur compounds (mineral or organic), turn intensely black, and have an unpleasant appearance.

All substances are objectionable which yield an alkaline effluent. This principle, strange as it may sound, at once excludes the commonest agent—lime. The reason is that putrefaction is more active in alkaline solutions than in such as are neutral or acid. In all experiments on the culture of the microscopic organisms, to which we have referred so often, and in all attempts to generate life from dead matter, alkalinity is a condition insisted upon. Now lime, if used not merely to neutralise some acid or acid salt, but, as a substantive precipitant, invariably renders the effluent water alkaline, and thus favours decomposition. Whether the lime is applied in a powder, as cream of lime, or as clear lime water, is merely a question of convenience and cost.

According to some recent investigations of Professor König and Dr. Boehmer (*Landwirth Jahrbücher*, 1885, vol. xiv., part 2, pp. 228–238), lime reduces the total organic matters suspended and dissolved in sewage by 33 per cent. in the tanks, and if the effluent be then allowed to flow over grass-land, by about 25 per cent. more, making thus a total of 58 per cent. But salts of alumina, with the aid of proper absorbents, are found capable of reducing the total organic impurities suspended and dissolved by 83 per cent. in the tanks alone, without flowing over grass at all.

A lime effluent is well known to be injurious to fish in any stream into which it penetrates. Of course, it will be said that the caustic lime present will soon be rendered inert, and be precipitated by the carbonic acid of the atmosphere. But if the supply of lime is continuous, a considerable tract of water may easily remain deadly to fish.

It will have been generally noticed that lime effluents and lime sewage deposits give off a very peculiar and most unpleasant odour. This, I suspect, is due to the volatilisation of some *ptomaine* (putrefaction alkaloid) present in the sewage.

The action of lime upon colouring matters in sewage and in industrial waste waters, discharged, *e.g.*, from tanneries and dye works, is very unsatisfactory. I have seen such waters easy to treat by other agents, but which, if mixed with lime water, were turned from a pale yellow to a dark and very permanent mahogany colour.

It is interesting to find that Professor E. Frankland,

in conjunction with Dr. Stevenson, has recently *recommended* lime as a precipitating agent for the Hendon sewage. The better to appreciate this advice, we turn to vol. i. of the first report of the " Royal Rivers Pollution Commission" of 1868, of which body Professor Frankland was the most active and prominent member.

In this document, published in 1870, we read:—

"(*a*). *Treatment with Lime.*—This process was, doubtless, first suggested by the ingenious operation devised by the late Dr. Clark, of Aberdeen, for softening certain hard waters.

" It has been applied to sewage upon an extensive scale at Tottenham, for the manufacture of Tottenham sewage guano; at Blackburn, and especially at Leicester, in the production of the so-called 'Leicester Bricks' (the name under which the manure was sold).

" In all these places the plan has been a conspicuous failure, whether as regards the manufacture of valuable manure, or the purification of the offensive liquid.

"We have witnessed the process at Blackburn, and on two occasions at Leicester, where it is still used, the machinery employed at the latter place being very perfect and efficient.

" At both places the method obviously failed in the purification of the sewage to such an extent as to render it admissible into a river. At Blackburn especially, the river below the outlet of the limed sewage was in a most offensive condition of putrefaction, our note, made at the time of our visit, being as follows : ' Horribly offensive, turbid, blackish stream, disengaging most offensive gases, with black masses of putrid mud floating on the surface.'"

A committee of most eminent chemists, not, perhaps,

remarkable for their special experience in the treatment of sewage, have recently devised and published a palliative process for the London sewage until the culminating extravagance of conveying it down to Thames Haven has been completed. These *savants* recommend the use per gallon of 3·7 grains of lime with one grain of copperas. It need scarcely be said that the effluent after the addition of this mixture will possess the defect of being alkaline. It may, indeed, to some extend "clarify," but it cannot "purify," the foul odour not being removed. In hot weather, therefore, it is recommended to add from 0·5 to 1·5 grain of manganate of soda, with half its weight of sulphuric acid. This formula will still leave the water alkaline, and in a good condition for the multiplication of micro-organisms.

It has been recently alleged by Dr. Percy Frankland that treatment with caustic lime (as in the Clarke process) removes bacteria from water to a considerable extent. The same authority has subsequently materially qualified this statement, informing us that after a few days the micro-organisms become more numerous than ever.

Lime is admitted to have a decomposing action upon the suspended organic impurities, in virtue of which they are in part rendered soluble.

Among the substances to be as far as possible avoided are further sulphates—compounds of sulphuric acid with various metals. The reason is that if brought into contact with moist carboniferous matter, such as sewage deposits, they may be gradually reduced to sulphides (sulphurets), which, in turn, if they meet with even the feeblest acids, liberate sulphuretted hydrogen. This result is easily noticed when copperas (ferrous sulphate, protosulphate of iron, or green vitriol) is used as a precipitating agent. The change is shown in this case by the intense black colour of the mass, which has been mentioned above.

Sulphate of lime (gypsum), or any mixture or combination by which sulphate of lime can be produced, is exceedingly objectionable. Although there is here no blackening, yet hydrogen sulphide (sulphuretted hydrogen) is given off in plenty. Instances are on record where men employed in mixing sewage deposits with gypsum to promote solidification have been rendered seriously ill.

Even sulphate of alumina—of which below—is probably one of the least desirable forms in which alumina can be introduced in the treatment of sewage.

Sulphurets (sulphides) are very rarely admissible. On no account should any soluble sulphide, or anything which may form such a sulphide, be allowed to pass into a river.

Hypochlorites, such as bleaching lime (commonly called chloride of lime), and the corresponding magnesia and soda compounds, must also be excluded. The late Royal Rivers Pollution Commission was fully justified in its protest against these substances as unfitting the water of rivers for almost every conceivable purpose, and especially rendering it deadly to fish. Mr. A. Anthony Nesbit, F.C.S., has shown within what narrow limits this deadly action takes place.

Salts of barium, such as barium chloride and baryta water, have been proposed for the treatment of sewage, and even for the improvement of drinking waters. They will, of course, remove from a water any sulphuric acid, free or combined, which may be present in solution, and will also precipitate carbonic and phosphoric acids. But they are more expensive than lime, the lowest figure for the native carbonate (Witherite) being 50s. per ton. Above all, they are very poisonous, and any portion passing out in the effluent or remaining in the deposit in a soluble state may work mischief both to animal and vegetable life. Salts of strontium, if they become cheaper, might be used with more safety.

Free acids, at least the hydrochloric (muriatic) and sulphuric, have been proposed. They are first to be run into the sewage, and then neutralised by the addition of an alkali or an alkaline earth, lime as the cheapest having probably been used. It is hard to see what good end can be reached in this manner, the alkali or alkaline earth undoing anything that the acid might have effected, and the final result being merely an addition of calcium sulphate or chloride (in older language, sulphate or muriate of lime) to the sewage.

Common salt has been used with no definite advantage. This result might be expected if we remember that this same salt, derived from urine, is one of the characteristic features of sewage, and that, if found in any water, it is considered *primâ facie* evidence of contamination with the excreta of animals.

Petroleum, coal tar, and similar products have also been recommended. They do not well mix with water; and, though they may at times mask an offensive smell, they cannot remove putrescent or putrescible matter. Being in themselves a nuisance in water, the Bill introduced last year by Mr. Walrond and Earl Percy very rightly proposed to make their emission into streams penal.

One of the strangest mixtures ever suggested is ground sulphur and turpentine! Being insoluble in water, and incapable of mixing with it, the action of this composition would be nil.

We come next to a class of systems which cannot fairly be condemned, but which are still not by any means free from objections. Here belong all the many processes which turn on the introduction of phosphates into the sewage. It is highly important that all the phosphoric acid originally existing in sewage should be removed, because such phosphoric acid remaining in the effluent water favours the multiplication of disease germs, and is, in addition, so much loss or waste withdrawn from the manure.

The phosphate processes are somewhat varied, the agent selected being either aluminium, iron, calcium (phosphate of lime), or magnesium phosphate. Of these the two first mentioned are generally dissolved in sulphuric or muriatic acid, avoiding excess. The solution is then run into the sewage and allowed to mix with it in a uniform manner. Lastly, milk of lime, or clear lime-water, is added, so as to neutralise the acid and cause the phosphate of alumina, or of iron, as the case may be, to be reprecipitated, occluding, as it goes down, more or less of the suspended and dissolved impurities. When phosphate of lime is the agent, it has been recommended to add to the sewage ordinary superphosphate, and then to precipitate with lime-water as above. Not a few inventors have proposed to remove at once phosphoric acid and ammonia from the sewage in the form of ammonium-magnesium phosphate, better known as the double phosphate of ammonia and magnesia.

All these processes are more suitable for the laboratory than for actual practice on the large scale. The large quantity of phosphoric acid added must be again removed, or there occurs not merely loss, but the multiplication of microscopic organisms is promoted. Gelatinous phosphate of lime, in the state in which it exists when freshly precipitated, decidedly promotes decomposition of an unsafe character. If superphosphate is introduced into the sewage sulphate of lime accompanies it, and if phosphates of alumina and of iron are used dissolved in sulphuric acid, sulphate of lime is thereby formed in the tanks. The objectionable character of this compound in sewage or in sewage mud needs no further demonstration. Further, in all these phosphatic processes the effluent must be kept on the alkaline side. This, except in the magnesia modification, is effected by means of lime. Consequently, all the objections against lime processes come here into force.

Further, a per cent. of phosphate of alumina rendered soluble by means of sulphuric acid, and a per cent. of soluble phosphate of lime, costs more than a per cent. of soluble alumina in sulphate of alumina, or in hydrated aluminium chloride. As for the phosphate of magnesia processes, it must be remembered that the double phosphate of magnesia and ammonia (ammonium-magnesium phosphate) is not a gelatinous mass like hydrated alumina, but has a granular texture, and is consequently much less adapted for occluding and entangling the organic impurities of sewage. Every chemist who has used the "magnesia process" for determining phosphoric acid must be aware that the conditions under which this acid can be entirely removed from a liquid are not such as can be produced in a sewage tank.

These considerations, combined with careful and prolonged experimentation on different scales, have driven me to the very unwelcome conclusion that the phosphate processes are not to be recommended for the treatment of sewage.

Having thus glanced at the principal agents which should not be used, we come to those which are more or less free from objection, and which may, therefore, be generally applied.

Foremost come the salts of aluminium. They are relatively cheap, inexhaustible, colourless, and harmless in any moderate proportion. As may be seen from their use as mordants, they have what is called a great "affinity" for organic matter. This holds good as well for sewage pollution, dissolved or suspended, as it does for the ordinary colours used by the dyer and the printer. If a solution of a salt of aluminium, say common alum or cake alum (sulphate of alumina), is thrown into a large quantity of water, it is decomposed. A basic sulphate, or as some call it, a subsulphate, is deposited, and combines with the major part of the impurities present in the water. This action is the more energetic if the water is slightly alkaline, as is usually the case with town sewage when free from industrial waste waters. The decomposition of the salt of aluminium and the precipitation of the impurities is also accelerated if it be added to the sewage *hot*, as proposed by Mr. W. C. Sillar.

It must now be asked, Which is the most suitable salt of aluminium to be applied in the treatment of sewage? Alum, though used in some of the earlier sewage processes, such as the original "A B C" process, is not to be recommended. In addition to its high price, it contains a considerable proportion of an alkaline sulphate (potassium or ammonium sulphates) which, without contributing in any degree or shape either to the purification of the sewage or to the agricultural value of the deposit obtained, make the effluent water "analyse worse." This is especially the case with ammonia-alum (double ammonium and aluminium sulphate), which causes the quantity of ammoniacal salts in sewage to be apparently *increased* by treatment. This fact (the apparent increase of ammonium salts) has been duly noted by certain official opponents of sewage precipitation, who have, at the same time, taken good care not to explain its cause.

Aluminium acetate, used by dyers and tissue-printers under the name of "red liquor," is a powerful precipitant, but its cost is prohibitive.

Much the same may be said of the nitrate, which, in addition, increases the proportion of nitrates in the water. The sulphate (cake alum, concentrated alum, or patent alum) is cheaper than alum, more rapidly soluble, and contains 15 per cent. of actual alumina, whilst potash alum contains only 10 and ammonia alum 11. If it contains a little iron, as in Spence's "alumino-ferric cake," no disadvantage is occasioned, but, if anything, rather an improvement.

Basic aluminium sulphates, where easily procurable, are, price for price, preferable to the ordinary sulphate, as they contain a larger amount of alumina, and are more readily decomposed in contact with the dissolved and suspended organic bodies; but, as it has been already remarked, sulphates are not to be selected except as a matter of necessity.

The best of the aluminium salts is the hydrated chloride, familiarly spoken of as muriate of alumina. Until recently this salt could not be prepared at a cost sufficiently moderate to admit of its being used in sewage treatment. Of late, however, a series of processes have been devised, suitable to different localities, by which a muriate of alumina, sufficiently pure for sewage purposes, can be produced at a very low price.

The aluminate of soda was proposed and patented as an agent for sewage treatment by A. J. Vassard, in 1871. More recently it has been patented, in a different combination, by F. Maxwell Lyte. Its precipitating power is indisputable, but the question of relative cost is somewhat doubtful. When used it must be accompanied by some acid salt, otherwise the effluent would, in ordinary cases, be rendered alkaline.

The aluminium salts have not merely the property of throwing down dead organic matter (dissolved or suspended) present in water, but they can, to a very great extent, remove disease germs. From time immemorial the Chinese, before using the very questionable water of their rivers for culinary purposes, have been in the habit of adding a pinch of alum to a tub of the water, and allowing it to subside. The French troops in Tonkin have adopted the same expedient, and have by this simple means almost entirely got rid of the endemic dysentery from which they previously were great sufferers.

Dr. Brautlecht has even proposed the use of alum as a means of detecting the microscopic germs present in water. He adds a few drops of a solution of alum to some of the suspected water contained in a test-tube, allows the precipitate to subside, decants off the clear, redissolves the sediment in a few drops of acetic acid, and searches for the organisms in the solution thus obtained.

It should, therefore, appear that the effluent water from any sewage process in which a salt of aluminium is the precipitating agent should be to a great extent freed from the disease germs in question.

An insoluble compound, or alleged compound, of alumina, has recently been recommended for sewage precipitation. This is the so-called carbonate of alumina. The chief authorities, such as Gmelin ("Handbook of Chemistry") and Watts ("Dictionary of Chemistry") do not admit the existence of such a body, though Muspratt (*Journal of Chemical Society*, ii., p. 210) alleges that he has obtained a true aluminium carbonate by precipitating a solution of alum with ammonium carbonate. The patentee precipitates, it would seem, solutions of cake alum with chalk (or soda-ash?), obtaining thus a mixture which, at any rate, contains sulphate of lime, undecomposed chalk, and either hydrated alumina (aluminium hydroxide), or the alleged carbonate, or possibly both. However this may be, I have been unable to find this mixture or compound at all more efficacious in purifying sewage, *if the same is neutral or alkaline*, than the quantity of cake alum from which it was originally obtained. In case of a strongly acid sewage it enables the use of lime water, etc., for neutralising the water to be dispensed with. This, however, is a distinction of little moment: if we add sulphate of alumina to acid sewage and neutralise with lime, the final result is the same as if chalk had been added to the sulphate of alumina before putting it into the sewage. The cost of the "carbonate of alumina" is, of course, somewhat higher than that of the cake alum from which it is made.

Carbonates of any kind, or indeed substances which can continue to give off gases after having been mixed with the sewage, are to be condemned. The bubbles of gas disturb the sediment, and cause minute particles to remain in suspension.

It is, of course, well known that freshly precipitated hydrated alumina, if well shaken up with the solutions of certain substances—*e.g.*, colouring matters—is capable of withdrawing them from solution. As far as the mere principle is concerned, this method would be applicable in sewage treatment, but in practice it is scarcely admissible, as the action is not sufficiently rapid and energetic, and the mechanical arrangements would prove too costly.

Soluble salts of manganese are excellent precipitants. Not only do they carry down organic substances, dissolved as well as suspended, but they destroy certain impurities by transferring to them a continual supply of oxygen from the atmosphere, and from the air dissolved in the water. This property is found more or less well marked in the salts of all metals which have two grades of oxidation. Any salt of a higher oxide in contact with organic matter is reduced to the lower oxide, and the oxygen which it gives off oxidises, or, in other words burns up, the organic impurities, whether dissolved or suspended. The lower oxide then takes up fresh oxygen from the air, is reconverted into a salt of the higher oxide, and repeats the former process. It is devoutly to be hoped that the opponents of the chemical treatment of sewage will not seek to deny that in such cases purification and clarification go hand in hand. Oxygen thus transferred to the particles of organic matter in sewage has a far more energetic action than the oxygen of air driven in by a pump.

A manganese process is at present in action in the town of Freiberg. Whether the manurial value of the deposit is reduced by the oxidising action just mentioned the writer has not been able to learn, though such a result seems likely.

Prior to the introduction of the Weldon process, the refuse of the chlorine stills at manufactories of bleaching powder was an admirable material for precipitation, either used alone or as an addition to the other precipitants. It was, in fact, patented in 1854, by J. A. Manning (No. 61). It is now no longer available, save in exceptional cases.

This may serve as an instance of the risks of basing an industrial process upon the employ of some waste product; either an improvement in the primary manufacture or its decline may render such waste product no longer available.

Sulphate of manganese, formed by heating black manganese ore with sulphuric acid, is too expensive an article for sewage purposes. But there has been recently proposed an interesting process by which this difficulty is got over. The inventor designs also to make a joint sulphate of manganese and iron.

Zinc, on account of its highly poisonous nature, cannot be applied in the treatment of sewage, though it has been repeatedly claimed.

Copper is costly, and is also poisonous, but, according to recent researches, so slightly so in small proportions that it is capable of being used along with other agents as a transferrer of oxygen.

Having thus glanced at the principal agents which

form combinations with organic impurities and carry them down, we come to a class of substances which act in a different manner, and may be very advantageously combined with the former. The bodies in question act upon impurities by absorption or occlusion. The substances of this kind are for the most part the same as those which are used in irrigation and filtration, but they are applied in an inverse manner. Instead of passing the sewage to be purified through clays, arable soils, grass-roots, etc., or through filter-beds made up of charcoals, coke, sand, peat, lignite, etc., we agitate these bodies in the sewage. Here is the advantage that fresh portions of the purifying material are continually brought into play, so that the annoyance of clogging or saturation never can arise.

As the best of these materials may be mentioned fatty clays, as free as possible from sand, grit, and especially from carbonates and sulphates of lime and magnesia. Such clay, in addition to its absorbing, purifying action, serves as ballast; it enables the sediment more readily to subside, and prevents it from being easily buoyed up again to the surface by any escape of gases.

Burnt clay, ground clinkers, etc., subside to the bottom, and, like sand, they occlude mere traces of the dissolved impurities. Their presence in the manure is objected to on good grounds.

Arable soil is as efficacious in precipitation as in irrigation. But a supply sufficient for use with the sewage of a town even of moderate size is not ordinarily to be procured.

This will at once appear if we reflect that such soil is rarely above a foot in depth, and that after its removal the land is rendered worthless for the farmer and the gardener for a time practically unlimited. It might, of course, after it has taken its part in a precipitation process, be brought back and spread upon the fields again. All chalky, lime or marl-soils are out of the question, on account of the carbonate of lime they present, which decomposes and wastes the metallic salts used in precipitation.

Coke has some absorbent powers, and it has been found by Dr. Percy Frankland very effectual in removing microscopic organisms from impure waters. It has been tried in a powdered state in precipitation processes, but, as it does not improve the resulting manure, the quantity in which it can be used is very limited.

Coal ashes have also been applied, but they have little to recommend them. They contain finely-divided silica no better than sand; alkalies, which impair the effluent, lime, magnesia and sulphur compounds, etc., none of which are desirable.

Peat, which in many parts of the United Kingdom is to be had without limit, is an excellent absorbing or occluding agent. It has also, as it is well known, a certain antiseptic action. It must, however, never be used in sewage or waste waters containing compounds of iron, or, of course, where iron is introduced in any shape into the precipitating mixture.

We come now to the various kinds of carbon : lignite, peat, and sea-weed charcoal, wood-charcoal, bone-black sawdust, charred by moistening with sulphuric or muriatic acid, the residual carbon from the manufacture of prussiate of potash, as also soot. In selecting any kind of carbon for sewage treatment certain points must be attended to. The carbon must be neither t dense nor too light. In the former case it settles at once to the bottom without having the time to act upon the dissolved impurities. If too light, it floats on the surface without coming properly into action, and giving the effluent water a very unsightly appearance. Suspended charcoal powder may be justly objected to in a stream, whether it is used for industrial or domestic purposes.

All charcoals employed must be thoroughly well burnt. If this is not the case they retain fatty and tarry matters, which repel the water and prevent it from penetrating into the pores of the charcoal.

On this account soot is very objectionable; it contains so much fatty matter as to float upon the surface of water without becoming wetted. It contains also a fair proportion of ammonia (1·65 per cent.), which, without aiding at all in the purification, causes the effluent to contain more ammonia, and, consequently, to analyze worse.

A very powerful absorbing agent is gelatinous silica. Without forming definite or permanent chemical combinations with the organic impurities, it entangles them, and carries them down. Its use in various combinations has been repeatedly patented.

Cement, Portland or Roman, has been repeatedly proposed as an agent for sewage precipitation. If it has any action at all (which is by no means demonstrated), it will be most likely as an absorbing and occluding substance. I am not aware that it has ever been even tried on a practical scale. That a material which sets or hardens in water can form the basis of a manure seems highly doubtful.

In addition to precipitants and absorbents, certain

substances are also occasionally used, which have, or are supposed to have, disinfecting, or, at least, deodorising, powers. Among these rank certain gases, such as chlorine, nitric oxide, sulphurous acid, atmospheric air, carbonic acid; further, carbolic and cresylic acids, carbolic sulphite, creosote, oils of coal and wood-tar, oil of turpentine, ethers, chloroform (!); also, chlorides of lime, magnesia and soda, sulphites and bisulphites, manganates and permanganates. Some of these bodies may be applied beneficially; others are utterly impracticable, being, *e.g.*, not capable of mixing with water. Some are mutually incompatible, the one undoing what the other has done. Of others, it may be strongly suspected that their chief effect is merely to mask or hide an evil odour, rather than to prevent or destroy it.

But of these substances generally it will be more convenient to speak in the chapter on deodoration.

If anything approaching to a thorough purification is intended, no one substance will effect the purpose. It is generally requisite to employ conjointly some absorbent body or bodies and a precipitating agent. The absorbents, which are best added to the sewage first, have the task of occluding and entangling in their pores offensive gases and noisome products existing in solution in the sewage. When these materials have become thoroughly incorporated with the sewage, and are saturated with the various nuisances, the precipitant, in the strict sense of the term, is then added. It forms insoluble compounds with much of the remaining dissolved matter, and coagulates both the suspended impurities and the fine particles of the absorbents (now charged with filth), and carries all to the bottom together. It is generally found that the offensive smell of the sewage is removed as soon as it has been properly mixed up with the absorbents, but that the liquid remains dull and turbid. On the other hand, the precipitant, if added alone, renders the sewage clear and bright, but does not in all cases entirely remove odours and colours. Both conjointly share the task of removing microscopic organisms, "germs," etc., from the sewage.

It will be then seen that the systematic conjoint action of precipitants and absorbents or occluders is needed, and that only those sewage processes can be trustworthy in which this co-operation is recognised.

It cannot be too clearly impressed upon the mind that a properly managed precipitation process is, or, at least, includes, *inverse irrigation*, and that this inversion gets rid of almost all the objections which can be raised against irrigation direct.

We must come now to the manner in which the chemical agents are brought to bear upon the sewage. The materials, dissolved, if soluble, or, if insoluble, ground to a pulp with water—which may be a portion of the sewage—are allowed to flow into and become incorporated with the sewage in a channel. This channel should be of such dimensions and construction that every drop of the sewage should have the opportunity to come in contact with the agents used. The channel then delivers the mixture into tanks, where ·the process of deposition or settlement takes place, whilst the clear water or effluent remains above.

This part of the process may be either intermittent or continuous. In the intermittent system we require a number of tanks, arranged side by side, each unconnected with, and independent of, the remaining. The treated sewage is first let flow into one of these tanks. When that one is full, the current is cut off and turned into the next tank, and so on, tank after tank being filled. As soon as the first tank is found to have settled, the clear liquid is allowed to run off by gravitation, if there is a sufficient fall, or is otherwise pumped off, with due precautions to prevent the mud at the bottom from being disturbed. The mud is then run off through distinct channels, or, if necessary, is pumped into a collecting reservoir, whence it is forwarded to the filter-presses, drying-floors, or other drying apparatus.

In the continuous process the treated sewage flows into the first of a series of tanks, all connected together, and from the last of which it passes constantly into the outfall channel. When any one of the tanks is found or judged to contain as much mud as is proper, it is cut off from the series, and pumped or run off, as in the intermittent system, whilst the sewage is let pass through the remainder of the set.

It is therefore necessary to have tank room enough to receive and deal with the day's flow even if one tank is temporarily disconnected. In this manner every tank is successively cleaned out. It will be, of course, evident that the tank into which the treated sewage *first* falls will require emptying much oftener than the rest of the set. Superabundant tank room is essential, as the bulk of the sewage varies greatly according to the weather. The construction of the tanks is a point of capital importance. The arrangement shown in the accompanying plan has been in use at the Aylesbury Sewage Works for some years, and, except as regards its size,

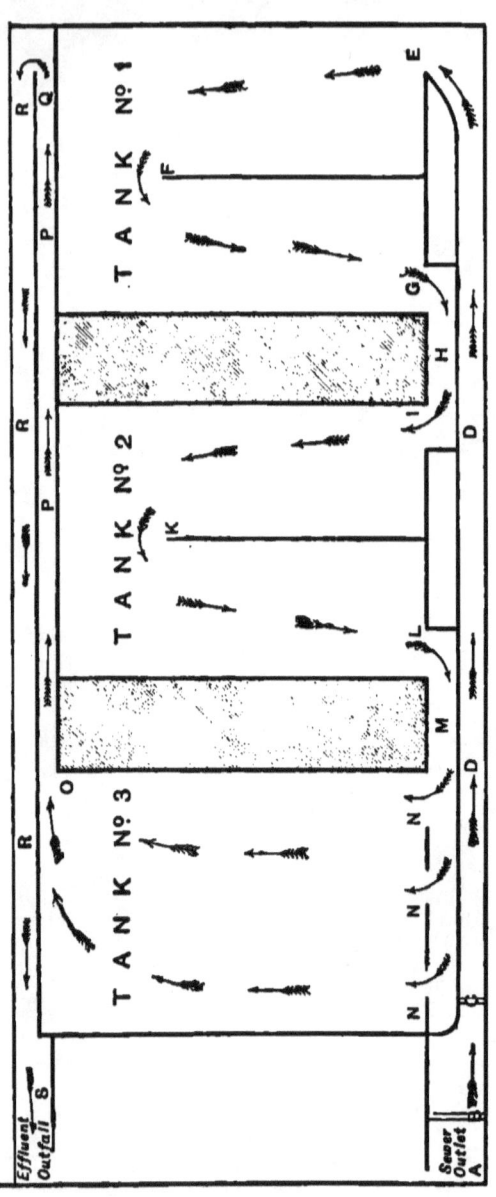

PLAN OF SEWAGE WORKS.

To face page 109.

has given general satisfaction. It is adapted for continuous working. The sewage of the town flows down to the works by gravitation, and is delivered from the sewer at the point marked A, where there is a grating to arrest rags, etc. Here it receives from the trough B a proper proportion of the purifying (absorbent or occlusive) mixture, which at once removes all offensive odour, and after flowing a few feet further, the precipitating agent (in the strict sense of the word) enters from the trough C. Where space is ample the distance between A and B might be increased with advantage. The sewage, having now received all the materials necessary for its purification, flows along the channel D to tank No. 1, entering it at the point marked E, and thence through the tank, passing round the partition-board F, through the outlet G, into the channel H. The water next passes into tank No. 2, at the inlet I, round the partition-board K, through the outlet L, into the channel M. It next passes through the three inlets N, N, N, into the tank No. 3, and, passing slowly through it, discharges at the outlet O, into the channel P, flowing thence round the partition Q, into the outer channel R, to the outfall S. Here it enters an open cutting about a quarter of a mile in length, and lined with concrete, finally venting into the river.

The mud, when it requires removal, is pumped out through pipes which pass from the front end of each tank beneath the channel D, and which can be connected as required with a powerful steam-pump. Any one of the tanks may be shut out of circuit if needed whilst the mud is being cleared away.

A specimen of well-arranged tanks for continuous working, but upon a much larger scale, may be seen at

Knostrop, Leeds. Here the level of the ground allows of the tanks being emptied by gravitation without any pumping, which is not possible at Aylesbury. The flow in sewage tanks must be as gentle as possible, so as not to hinder the subsidence of the precipitate, or disturb that which has already settled. To this end the fall from B to S, and especially from E to S, should be minimized. In the arrangements for passing the sewage from tank to tank, or from tank to channel, eddies or whirlpools should be most carefully guarded against; they *bore* the mud from the bottom up to the surface.

Speaking within compass, I venture to say that ill-planned tanks, which require an extra allowance of chemicals to force precipitation, may lead to a waste of 20 per cent., and then not give a satisfactory result.

The proportions of the ingredients to each other, and to the impurities to be dealt with, cannot be usefully modified beyond a certain point.

For a large town, where the number of tanks must necessarily be considerable, they are best placed in a double row. If arranged in a single file, a long narrow slip of land is needed, which is not always easily to be had. There is also trouble—at least where gravitation is not available—with the appliances pumping out the mud; these must, of course, be capable of acting in every tank.

As it has been already said, ample, superabundant tank room is desirable. In all towns where a double system of sewerage is not in action, we have to keep in view the possibility of storm-water. The sewage of a town, which in fair weather averages 250,000 gallons daily, may by a sudden heavy thunder-storm be swelled to a million gallons or upwards. If such a flood is allowed to force its way into a set of small tanks, the deposit is washed up again, and swept down into the river, giving rise to not unjustifiable complaints. On the other hand, it is not advisable to let the excess of water sweep by untreated. The sewers, after prolonged dry weather, will be found coated with dry fæcal matters in very foul states of putrefaction. All this is brought down by a sudden storm, and is certainly harder to treat than normal sewage.

Some engineers, in laying out sewage works, have proposed to roof the tanks over. This is a capital mistake. The treater cannot know whether the process is working satisfactorily or not without he can walk round the tanks and see them in different lights. Further, the free contact with the air and the access of light assist no little in purifying the sewage. If the tanks are found to give off a bad smell, it is proof that the process is not being satisfactorily worked, or that it is essentially deficient, and needs to be superseded.

The tanks shown in the plan are considerably deeper at the inlet end than at the outlet end. In some works the depth is alike at both ends, and in others again, the inflow end is made the shallowest. Decisive experiments on the comparative results of these three arrangements, all other circumstances being kept the same, are still wanting.

It is also to be regretted that engineers—to whom it evidently belongs—have never fully solved the following problem : For a given flow of treated sewage, what should be the shape, size, and arrangement of tanks, so that settlement may take place the most quickly and completely?

The importance of this question has on several occasions painfully forced itself upon the writer's attention.

A great mistake is, as it has been proposed by some inventors, to have two or more successive precipitations, the sewage being treated in one tank with some given material, and in another with a second agent, etc. Such systems of necessity lead to a duplication not merely of tanks and channels, but of grinding and mixing machinery, of mud-pumps, ducts and reservoirs for different kinds of sediments, as well as to an increase of the mechanical power and of the labour needed. Now, is there any set-off against all this extra outlay?

Whatever process be adopted, and whether the working be continuous or intermittent, among the most essential requisites are CLEAN TANKS. In their default the best process in the most skilful hands will in a great measure fail. The reason for this is not far to seek. The deposit formed at the bottom of the tanks, if left for many days in contact with water, begins to undergo change. Matter which had been rendered insoluble tends to decompose, and fouls the water above. Bubbles of gas—chiefly hydro-carbons—are liberated, and buoy up lumps of the mud to the surface. Here they burst, and diffuse the finely divided mud through the water. These changes take place especially if the barometer is low and the temperature high. It is to be remarked that the sewage mud from a proper precipitation process, when freed from water, does not give off any bad odour, or appear to undergo any rapid changes.

Where a very high grade of purification is aimed at, the water, as it passes away from the last tank, may be aërated. This can be effected in various ways: where space allows, it may be permitted to flow in an open, shallow channel for some distance before being discharged into a stream. Or along the outflow channel there may be laid, below the water level, pipes perforated with a number of fine holes, through which air is forced by any available power. Or the water may be made, if the ground allows, to form a series of cascades or rapids. Or, before being discharged, it may be filtered over a bed of coke, lignite, peat, blackstone carbon, or other suitable material. Or, lastly, it may be utilised for irrigating any plot of land which may be the better for an increased supply of moisture.

It has been already mentioned that where there is scope for such an arrangement it is advisable to let the sewage average itself before treatment in a collecting tank.

At some sewage works the liquid before treatment passes through revolving wire screens of different grades of fineness, no less than 4-horse power being in one place consumed in setting this machinery in motion. This seems a costly and a needless complication. If the sewage, before entering the channel where the treatment takes place, is allowed to pass through a grating of iron rods, laid in a direction sloping away from the current, all coarse suspended matters, corks, rags, dead rats, lumps of solid excreta, etc., are arrested, and may be easily cleared away with the stroke of a rake and thrown into a wheelbarrow for removal. With this grating and the occasional attention of a labourer, this object is as well effected as with costly machinery and a considerable amount of steam power.

It is desirable that the grinding mills and the mixing pits in which the ground materials are kept ready for use should be close over the sewer mouth. Thus the foreman who conducts the treatment can watch the character of the sewage as it enters the channel, and on the appearance of any change in its quantity or quality can, without stirring a step, by simply reaching his hand to a cock or a valve, make the corresponding change in the quantities and proportions of the mixture used.

To control the process the following very simple method may be employed: Suppose that we are using two ingredients, a and b. We take four small hydrometer glasses and fill them to a known height with the sewage *after* it has received the mixture. No. 1 glass we leave as it is. To No. 2 we add a little more of a (a solution of which is kept ready at hand for the purpose); to No. 3 a little more of b, which is also kept ready; and to No. 4 a little more raw sewage. We then observe which of the four glasses is the best, *i.e.*, which goes down with the cleanest, boldest flakes, leaving clear, colourless, inodorous water above and between. If No. 1 is the best, the treatment is continued as it was; if No. 2 is seen to be an improvement, we increase the proportion of a added; if No. 3 has the advantage, we increase in like manner the dose of b. And if No. 4, to which more raw sewage has been added, is the best, we see that we have been using too much material, and we reduce the quantity accordingly.

It is easy to sneer at this method as a "rule of thumb" procedure,—an expression often used in a very "rule of thumb"-like manner. But considering that in many towns the character of the sewage may, and often actually does, undergo a decisive change within a few minutes, we shall see that no process of chemical analysis is sufficiently rapid. There are other methods by which the treater may be guided. He may from time to time dip up a hydrometer glass full of the treated sewage from different points of the channel D (see plan), and set them aside in a good light to observe how they settle. He should occasionally walk round his tanks and inspect them from different points of view as regards the light. If any process is acting well, the effluent water, when in bulk, appears of a peculiar bluish tinge. This indicates not merely "clarification," but "purification," since this blue tint, according to the researches of Mr. W. Crookes, F.R.S., is the more decided the freer a water is from dissolved organic matter.

The treater must learn, by careful, intelligent observation, what modifications the sewage he has to deal with generally undergoes; at what hours, on what days, or under what conditions of weather, etc., these changes come on, and how they are all to be met.

It must be remarked that the specific gravity of a sample of sewage gives no clue to the quantity or quality of the treatment it will require. Mineral substances in solution, which raise the specific gravity of a sewage, may cause it to require *less* absorbing and precipitating matter than a sample specifically lighter, because richer in organic matter.

Sewage is more readily precipitated if warm than when cold. There is also a decided advantage in using the precipitating agents hot, as proposed in a recent patent.

It has been suggested, by way of economy, to use the deposit over again as an occluding or absorbing agent with a fresh dose of a metallic salt, and with or without a recruit of carbonaceous or earthy matters. The experiments made in this direction have not been remarkably successful. The first time of re-using, the effluent was found, perhaps, nearly as good as one obtained with fresh materials; the second and third times gave decidedly worse results, and on trying a fourth and fifth re-usal the effluent was bad in colour and odour, and showed on analysis little improvement if compared with raw sewage. This result cannot be termed unexpected. The occluding power of carbon, in its various forms, and that of clay, arable soils, etc., cannot be supposed infinite, and when saturated they cannot take up more. If they are then again and again brought into contact with fresh portions of polluted liquids, the probability is that, instead of taking up any more putrescent matter, they may part with some that they had already absorbed.

Nor is the prospect of economy very clear even if it were found possible to re-use the deposit once or twice. In that case, to work systematically, and to know what is being done, it will be necessary to keep distinct—a, the deposit obtained with fresh materials; b, the deposit obtained on once re-using; and c, the deposit obtained on twice re-using. To do this on a practical scale requires considerable trouble and expense, since each kind of mud, on cleansing out the tanks, requires to be drawn off into a separate receptacle, and the lots which are to be re-used have to be pumped back into the mixing pits.

It may not be useless to consider some of the causes which have raised so strong a feeling against the chemical treatment of sewage—a prejudice which, though abandoned by the majority of observers in deference to facts, still lingers in official quarters.

In the first place, we must admit that a multitude of sewage processes have been devised without the lights either of theory or of practice. Instead of selecting materials which might mutually support and supplement each other, inventors seem mainly to have looked out for articles which were cheap, or entirely worthless, and heaped them together without any definite notion of the part which they were separately and collectively to play. This alone can account for the recommendation of such bodies as coal-ashes, soot, salt, gypsum, etc., which in almost every case would do more harm than good. Very often we see, especially in the older specifications, materials given as alternatives whose action, if any, must be evidently quite dissimilar the one to the other.

Not uncommonly, to furnish the basis for a patent, some well-known and often-used agent was proposed, and with it, to give an air of novelty, some useless or even mischievous matter.

In addition to "bogus" patents there was "bogus" working. A municipal authority, or a company, or a contractor—it does not matter which—might be nominally and ostensibly working according to some process which has much to be said in its favour. But in daily practice the most important agent would be left out entirely, or else used merely in quantities sufficient to swear to. If some important personage or deputation was expected, then the omitted article was brought into use, and the good result produced was triumphantly exhibited as that of the ordinary, normal working. Further, there is no obligation on a man who contracts to purify sewage to work under any patent at all. He may legally and equitably make use of any patent which has expired and become public property. Or he may make a combination of such methods. Or he may have a secret process of his own. But if a man thus acting professes to be working under a patent which has expired, or which has never existed, he is sailing under false colours.

Another great misfortune has been the exaggerated expectations entertained and fostered by the earlier sewage purifiers concerning the commercial value of sewage. From calculations, perfectly correct in themselves, of the waste incurred day by day in the form of ammonia and phosphoric acid poured down the sewers, the inference was drawn that the whole of this value was recoverable. Hence, certain municipalities claimed payment for their sewage, and individuals, syndicates, and companies were found sanguine enough to offer such payments, and still hope to cover all the working expenses, and secure a fair profit on the capital invested.

From this error, which, be it remarked, was merely one of the delusions common in the early days of any new undertaking—*e.g.*, electric lighting—all persons concerned have now recovered. Whilst insisting, as will be explained in a future chapter, that a very considerable amount of plant food—*i.e.*, manurial matter—can be obtained from sewage, either by irrigation or by judicious chemical treatment, we must clearly admit that the value thus realised is not sufficient to pay working expenses, and leave such a margin as will remunerate capitalists who take the matter in hand. If a firm or a company undertake to guarantee the treatment of the sewage of any town, and furnish security for the due execution of the undertaking, the town must expect to pay a reasonable subsidy.

Concerning opinions which are the mere result of prejudice, and which their authors dare not retract for fear of owning themselves in the wrong, there is no need to enlarge.

CHAPTER X.

DEODORISING.

THE title which, for want of a better, is given to this chapter, is not, perhaps, as intelligible as might be desired. In irrigation and precipitation processes one of the main objects kept in view is to deodorise the sewage. But the processes about to be here discussed do not seek to remove the impurities from sewage or other foul waters, but merely to prevent them from being immediately offensive. They are, in short, palliative rather than either preventive or curative. They derive at present an exceptional and transitory importance from the attempts made by the Metropolitan Board of Works to deal with the present state of the Thames, as arising from the practical failure of the Bazalgette system, without having been compelled to own themselves mistaken.

In order to form an opinion concerning the promise and potency which are in deodorising processes, let us take one which was actually used by the Metropolitan Board of Works. Chloride of lime (bleaching powder) was, as it is said, put into the sewage not in any suitable precipitation tanks, but into the river itself, after it had been polluted by the discharge of sewage at Barking Creek and Crossness. The deodorisation, if it was anything more than mere masking, could but be temporary.

The active chlorine must soon escape, and the organic matters present could not be destroyed or rendered permanently inactive by any proportion of the bleaching agent that could possibly be applied. Were this point actually reached the crews of vessels on the river, and even all persons living or working near the margin, would be subject to great inconvenience, if not to actual danger. Many kinds of cargoes and all the metal fittings of ships would be injured or corroded. Fish, if present, would be at once destroyed. The precipitation of the various pollutions could not be notably greater than that of a lime process, and the matter precipitated, be it more or less, would be deposited at the bottom of the river. Here it could not be regularly removed, as would be done in a precipitation tank. Nor would matters be greatly improved if chloride of lime were applied to the sewage in tanks. To treat only a part of the sewage in this manner, leaving the rest in its original nastiness, could only be regarded as playing with the question. And if we suppose the whole of the sewage of London, say 170 million gallons, daily collected in tanks and there treated with chloride of lime, a nuisance would be generated far greater than any of those which the Alkali Act was intended to combat.

But chloride of lime soon departed, though not in the odour of sanctity, and was succeeded by manganate of soda. The alkaline manganates and permanganates, and, indeed, various other compounds of manganese, have long been used for transferring oxygen to organic matters in solution, or, in other words, for burning them up. Thus, W. C. Sillar, G. R. Sillar, and G. W. Wigner, in a patent, No. 1,954, of 1868, claim manganate of potash as one of the ingredients in the original " A B C " mixture. Whether it was actually used in this combination I am unable to say; but the manganates and permanganates have some serious drawbacks as sanitary agents. They do not actually destroy the organic matters dissolved in water unless introduced in relatively large quantities and assisted by heat. Chemists who have been accustomed to determine organic matter in drinking waters by the aid of the permanganate process in any of its modifications, will be slow to believe that, in the proportions introduced by the Metropolitan Board of Works in their operations at Crossness, it can have had any marked effect. The permanganate has certainly not been added to the entire volume of polluted waters, and anything less is trifling. Experiments? But what is the use of experimenting on points already decided? That permanganates, unless accompanied by a quantity of sulphuric or other acid sufficient to neutralise all the alkali, must leave the sewage in an alkaline state, needs no demonstration. Their inability to destroy micro-organisms is well known. I have seen animalculæ large enough to be distinguished with the naked eye surviving for an entire day in water to which potassium permanganate had been added in excess. How long they would have survived I am unable to say, as a laboratory boy took it upon himself to empty out the contents of the vessel.

This is, perhaps, the proper place to notice certain attempts, or at least proposals, to deal with sewage by forcing into it common air or antiseptic gases. Aëration, as supplemental to precipitation, has already been noticed, and where a very high degree of excellence is required, it is at least worthy of trial. But aëration has also been proposed as a substantive process for grappling either with the crude sewage or with the sewage merely freed from its solid impurities. Thus James Bannehr (No. 2,918, A.D. 1867) treats the liquid portion of sewage with an acid to fix the ammonia, passes through the chamber containing the mass under treatment a current of air, which may be warmed and "charged with a current of electricity or of carbonic acid." Ferrar Fenton (No. 1,897, A.D. 1871) passes through the sewage "a blast of atmospheric air." E. Hills and B. Biggs (No. 3,464, A.D. 1872), after having run the sewage into an air-tight tank and added sufficient lime to set free the ammonia, force atmospheric air through such sewage into another tank, where it encounters sulphurous acid to arrest ammonia and decompose sulphuretted hydrogen. R. S. Symington (No. 912, A.D. 1873), after filtering the sewage upwards through ashes, exposes it to the action of the atmosphere "by falling in a broken manner through a sufficient height before passing through the last filtering tank." G. Rydill (No. 399, A.D. 1875), after filtering foul water and treating it with caustic soda or lime, "forces air into it from perforated pipes arranged in a tank and connected to a blower." Lastly, a Mr. James, of Tottenham, recently patented an arrangement for the treatment of sewage by aëration, and has called attention to his system in a circular which contains some rather hazardous assertions. He proposes to run raw sewage into an air-tight tank, at the bottom of which lie a number of perforated iron pipes, into which air is forced by some suitable mechanical arrangement. The air, after having bubbled through the sewage, escapes through two pipes, one of which is supposed to convey the sewage gases, and the other the air deprived of its oxygen, into a chimney 150 feet in height, whence it escapes into the atmosphere. The circular makes no mention of any arrangement for forcing the sewage gases, possibly charged with microbia, through a furnace.

How long this treatment has to be continued before a tankful of sewage is purified, the inventor does not say. This is to be regretted, since without this point is ascertained, approximately at least, neither the first cost of the plant for a given flow nor the subsequent working expenses can be calculated. I must avow my suspicions that the time required for thus purifying raw, undiluted sewage by the mere contact of air will prove very long. In rivers which are to some extent polluted with sewage there are, as it is shown in the chapter on self-purification, certain agencies at work which in the treatment of undiluted sewage in an air-tight tank are out of the question. Nor is the high chimney a satisfactory arrangement. High chimneys have been found to do little towards mitigating the nuisance from the discharge of sulphurous or hydrochloric acid gases, chlorine, coal-smoke, etc., into the air. The sulphuretted hydrogen, and the liquid and solid impurities which must be mechanically carried aloft by the ascending current, will be brought down again to the earth by rain, and will be introduced where they are not wanted.

Nor must it be forgotten that in this process, as in Bazalgettism and "intermittent downward filtration"— but unlike irrigation and precipitation—the entire cost is incurred in pure waste. No fraction of the plant-food contained in the sewage is returned to the soil, but all is hurried out to the sea. Hence this process, surely, is greatly to be deprecated.

It must be remembered that Mr. James proposes to

lay his perforated air-pipes, not merely in special tanks, but also along the course of the sewer. Here, we presume, the sewage gases will have to escape, not by way of a chimney into the upper regions of the atmosphere, but through the sewer-grids and ventilation holes into the streets.

Some inventors have proposed carbonic acid, either as an auxiliary or as a sole agent, in the treatment of sewage. There is nothing intrinsically absurd in this idea. Carbonic acid has a certain antiseptic power, as it has been lately re-discovered, and as a sequel to a lime process it might be useful.

Nitrous oxide, chlorine, sulphurous acid, and hydrochloric acid, have been patented as agents—sole or merely adjunct—in the purification of sewage. In very rich animal liquids containing little but urine, and especially blood, like the drainage from slaughter-houses, treatment with chlorine is advantageous. With blood its behaviour is remarkable; the blood quickly sets into a loose, dry, granular mass, incapable of offensive changes, and well adapted for distant transport. For ordinary town sewage agents of this class are inconvenient, and far too costly and troublesome in their application.

CHAPTER XI.

DESTRUCTION.

THE organic matter contained in sewage is essentially manurial, including, as it does, all the nitrogen eliminated from the bodies of men and other animals, as well as the phosphoric acid and potash. Hence every process by which this manurial matter is applied, or sought to be applied, not to the fields and gardens where it may serve as plant-food, but to any other purpose, or is, in conventional language, destroyed, wasted, may be termed "destructive." Some of the processes already discussed might fairly share this name, such as the Bazalgette system, the aëration process of Mr. James, and, indeed, "intermittent downward filtration" itself.

But, as types of destruction, we may take the processes patented by the late General Scott (Nos. 849, A.D. 1872, and 296, A.D. 1873). It is commonly said that by these processes he converts the sewage matters into cement. This is not a correct way of looking at the matter. He adds to the sewage certain substances from which cement can be made in the absence of sewage. By these added substances the impurities present are precipitated and entangled (in part, at least), and when the deposit is calcined the organic portion of such impurities is destroyed. But that the sewage matters contribute anything to the value of the cement, or that the cement is in any way improved by sewage having been implicated in its formation, is not proven. The case is like the far-famed "pebble-soup" of the mendicant friar, or the petroleum soap of the American "crank."

Concerning the quality of the cement thus obtained practical men have given their opinions, and, so far as the writer learns, these are not flattering.

The charge of destruction brought against this process, and every other where it is sought to convert sewage matters into cement, ballast, bricks, or other non-manurial products, is based upon the following facts: The free ammonia, and that present in ammoniacal salts, will be expelled by the excess of lime added, and will escape in the gaseous form ; or, if in part entangled in the precipitate, it must be driven off in waste during calcination. The organic nitrogen will, in part, be precipitated, and will also be destroyed during calcination. The phosphoric acid, in as far as it is rendered insoluble by the excess of lime, will also be withdrawn from circulation, being not indeed destroyed, but rendered useless. Now, the supply of phosphoric acid and of combined nitrogen in the world is not unlimited, and as these substances are the scarcest items of plant-food, their destruction or misapplication is a serious crime against humanity in general, and ought in every way to be discountenanced.

CHAPTER XII.

PROMISCUOUS METHODS.

WE consider here a number of systems of sewage treatment which cannot be placed under any general head. Foremost may come distillation. It has been repeatedly proposed in patents, to collect the sewage in suitable stills, with or without the previous addition of agents to promote the liberation of ammonia, and to distil. The ammonia, driven off by the heat is to be passed into suitable vessels charged with sulphuric acid, and thus absorbed and rendered marketable as sulphate of ammonia. The methods for carrying out this process vary in minor details. In some cases the raw sewage is to be distilled; in others the effluent from some precipitation process; and in others, again, the sewage-mud, if from any cause it is not applicable as manure, is to be submitted to destructive distillation. One circumstance first pointed out by Professor J. A. Wanklyn is much in favour of these distillation processes. That is, the ammonia is given off almost entirely at an early stage of the distillation. In some of these processes the separation of the ammonia from the sewage is promoted by driving through it a current of hot air, or steam.

The great difficulties in the way of these distillation processes are—I. The enormous size, or number, of the stills needed for treating the sewage of a large town, and the question: 2. What is to be done with the residual liquor? For the ammoniacal salts in the sewage, which alone are extracted by this process, if perhaps the most saleable, are not the most dangerous or noisome constituents of sewage. The residual liquor will contain the organic carbon and organic nitrogen, the nitrates, nitrites and phosphates, and must be disposed of either by irrigation, filtration, or by a precipitation-absorption process.

One system of distillation escapes some of these objections, by means of extreme complication. The sewage as it enters the works passes through a series of revolving gratings, becoming successively finer and finer. In this manner are arrested corks, lumps of soap, fatty matters, hair, rags, etc., and these materials are said to be capable of profitable utilization. Whether they can ever be cleansed and disinfected so as to make them again fit for coming in contact with the human person or with articles of human consumption, is, perhaps, open to doubt.

After escaping from the revolving screens the sewage runs into tanks, where it is precipitated by the lime process.

The operations now become twofold: *a.* The sewage mud or sediment is collected, dried, and submitted to destructive distillation in retorts similar to those used in the gas manufacture. The nitrogen present in organic combination will be driven off, chiefly as ammonia, and may be collected by means of an acid scrubber, yielding sulphate of ammonia.

The residue in the retorts, which will retain the phosphoric acid, may be sold as manure, provided that anyone will buy it.

b. We return now to the effluent from the lime precipitation. This liquid is passed into large receivers, from which the air can be almost entirely exhausted. The effluent gives off its ammonia, which is pumped into sulphuric acid, thus forming a second lot of sulphate of ammonia. It is evident that this process must require a costly plant. It is not evident that the lime effluent, after the removal of its ammonia will be fit for admission into a river without further outlay in the way of filtration or irrigation. It will, of course, be alkaline.

Freezing by means of artificial cold has been proposed for sewage treatment. There still prevails a notion in some quarters that water, in freezing, rejects its impurities. Hence the idea was conceived that by applying cold to sewage its pollutions might be concentrated in the portion remaining liquid, whilst the ice formed would be pure, and might either be thrown into any river, or even sold for any purposes to which ice is applicable. This entire process is founded on a mistake. The suspended pollutions in foul waters are entangled in ice during freezing just as are leaves, straw, etc., in a pond. Nor is it very different with soluble impurities, Ice from impure water has been found to contain organic nitrogen and carbon, and has been recognised as dangerous to health. Lastly, microscopic organisms are not killed by freezing, nor even by much lower temperatures, and resume their activity when the ice thaws.

It must further be borne in mind that freezing sewage on the large scale would be a very expensive process.

One inventor (No. 2,997, A.D. 1871) proposes to reduce the temperature of the air above the tank "by allowing liquid carbonic acid to evaporate into it."

K

Cheaper methods of obtaining cold are now available, but even by the most economical process the partial freezing of the sewage of a town of moderate size would be a most costly undertaking. Some advocates of freezing recommend that precipitating agents should be added to the sewage before or after refrigeration, a step which would improve the result, but, at the same time, swell the working cost.

Another agent recommended is electricity, generally in combination with other processes or agencies. Thus C. F. Kirkman (No. 2,653, A.D. 1870) proposes a filtration process, but suggests that on its way to the filters the liquid should pass through a receptacle in which are a number of zinc and copper plates, by which arrangement "a continuous current of electricity is made to pass through the sewage water, and will materially aid in freeing it from its manurial properties."

F. H. Atkins (No. 556, A.D. 1873) also proposes to apply "galvanic, magnetic, or electrical action to filtering apparatus, reservoirs, or tanks, for the purpose of precipitating organic and inorganic matters or impurities held in suspension or solution in the water or other liquids." He arranges plates in the tanks, and passes magnetic or electric currents through the liquid by using such plates as electrodes.

E. H. C. Monckton (No 265, A.D. 1874) proposes to use "electrified channels" for purifying sewage, or to drive ozonised water (ozonised, it would seem, by electric action) into sewage. He likewise proposes to employ windmills to generate electricity, which may be used for purifying sewage.

On these and other similar processes it may be remarked that magnetism has not been found to have any distinct power of occasioning chemical decomposition, and that special experiments have yielded no definite results. With electric currents the case is different. From experiments on a small scale, the writer is disposed to conclude that imperfect effluents from any kind of process might thus be brought up to a high degree of purity. On the question of working cost it is difficult to decide, but probably where there exists a fall of water, or other inexpensive means of turning a dynamo-electric machine, the process might be practicable.

Contact with bundles of iron wire has been suggested. I am not aware that it has ever been actually tried, nor does its manner of action seem intelligible. It is not said that the wire is to be formed into a filter or strainer.

It may seem surprising that desiccation, in other words, simple drying up, has been recommended as a method of dealing with sewage. The writer is not aware that this process has been tried or even suggested for raw sewage. But it has been proposed to get rid of the "solids," to expel the ammonia by the addition of lime, and then receive it in sulphuric acid. The water, after being thus deprived of its suspended matter and ammonia, was to be evaporated to dryness in covered vessels, so arranged that the fumes should be carried through a furnace and up a chimney. The great expense of treating sewage by any process in closed tanks makes it unnecessary for us to examine this proposal any further.

Here also must be mentioned a system in which cement is used as a precipitating or absorbing agent W. H. Hughan (No. 2,883, A.D. 1868) adds powdered cement to sewage *in the sewers*, and then delivers it into a tank "where precipitation takes place." The sludge is then mixed with more cement, until the required degree of consistency is obtained. Here it would seem that the cement is to be added until the whole is brought to the consistency of ordinary mortar. Portland or Roman cement is used, but along with it 1 part of copperas to 6 parts of cement, or the material may be calcined. How such a deposit can serve as a manure is not easily understood, if we consider that the cements above-mentioned contain from 54 to 56 per cent. of lime, from 21 to 36 of alumina, and from 8 to 15 per cent. of oxide of iron, and that they further quickly "set" to a hard mass with water.

"Screening," as distinct from filtration, has had its defenders, especially in the earlier days of the sewage question. Certain inventors arranged very elaborate screens of gratings, wire gauze, etc., and were evidently of opinion that when the visible suspended impurities were removed the water might be safely admitted into a river. It is plain that the purification thus effected, unlike that of an earth or a carbon filter, could be merely mechanical, and that the more dangerous part of the impurities remain.

A peculiar method of treating, or rather of partially utilizing, sewage has been proposed by A. M. Graham (No. 4,791, A.D. 1876). This inventor precipitates the sewage, first rendering it alkaline, by any suitable process, preferably by the use of a salt of alumina, iron, or manganese. The sediment thrown down is treated with sulphuric or hydrochloric acid, and heated in a retort until the fatty acids, water, etc., are driven off. The fatty acids are then collected from the water.

It is a fact that a certain quantity of fatty acids can thus be recovered, but it is very doubtful whether the product recovered could cover the expense of the process. Large consumers of soaps generally collect the waste lyes separately and recover the fatty matter.

Perhaps this is the place to notice the so-called "Metropolitan system," that, to wit, which the Metropolitan Board of Works seem about to apply to the sewage of London. The first stage is precipitation with a small quantity of lime ($3\frac{1}{2}$ grs. per gallon) and copperas. The effluent is then to be dosed with a mixture of the permanganate of soda and sulphuric acid. Next follows aëration, air being forced through the water in a multitude of fine bubbles. Lastly, the deposit is to be put on board ship, carried out and sunk in the North Sea! This will doubtless remind the reader of Dr. Abernethy's celebrated receipt for dressing cucumbers. It need only be pointed out that in this process the entire expenditure is incurred in pure waste, no value being recovered from the sewage.

CHAPTER XIII.

SELF-PURIFICATION—CELLULAR CHEMICAL TREATMENT.

THIS subject must be viewed in two different lights, according as we refer to tanks filled with sewage, or to rivers more or less polluted with the sewage of towns and with industrial waste waters. That sewage left to itself in tanks or pools will become spontaneously purified must, *practically speaking*, be denied. If it be allowed to stand exposed to air and light for a time longer or shorter according as its quantity is larger or smaller, and its foulness more or less intense, and if the access of fresh supplies of putrescible matter is completely excluded, there comes, indeed, a point when the most palpably offensive phases of putrefaction are at end. Bubbles of gas are given off rarely or not at all, and fetid matter is no longer deposited. The water may even have become limpid and colourless, but it would be very hazardous to regard it as pure. The pool will have become a nursery for micro-organisms, possibly including disease germs, and a sediment of foul mud will remain at the bottom. Moreover, during the whole time of exposure the pool or tank will have been a serious nuisance to all persons living near or even passing by it. On these grounds, and also because the time and the space required for this imperfect self-purification are both excessive, I am not aware that anyone now—in Britain, at least,—seriously proposes to deal with the sewage of a town on this principle. A number of patents were at one time taken out for dealing with sewage by simple subsidence without the aid of any agent, occluding, precipitating, or deodorising. The clear liquid was then to be let off into a river or into the sea, and the deposit was to be dealt with in various manners, as if it were either the most valuable or the most dangerous portion of the sewage. Some of the patentees had, indeed, the discretion to suggest that the gases and fumes given off from the subsidence tanks should be passed through a furnace or up a high chimney-stalk.

Treatment of sewage by simple subsidence was once tried at Knostrop, near Leeds, at the suggestion of an official. When, a few days afterwards, the Sanitary Committee of the Town Council went to see—or smell —not one of them, it is said, ventured within some yards of the tank. But all experiments made formerly in this direction, and all opinions expressed as to the practicability of subsidence, may be passed over in charitable silence.

For waste waters, not containing any sewage, in the strict sense of the term, but rich in carbo-hydrates, fermentation is much employed on the Continent. Urine, and sometimes lime, or any other agent deemed suitable, may be added to quicken fermentation. Thus refuse waters from sugar works have first been allowed to subside so as to remove the coarser suspended matters. Then followed treatment with precipitating agents (sometimes spoken of abroad as molecular chemical treatment), and the effluent was next brought into fermentation in separate tanks. Lastly, the water was used for irrigation. Concerning this very circumstantial and necessarily costly process, it must be remarked that precipitating agents, so far as it is at present known, take very little effect upon sugar existing in solution. But the preliminary subsidence process, prior to the addition of the precipitants, is utterly needless, a mere waste of time and of plant. It will be at once understood that the fermentation process (cellular treatment) cannot be carried out in the same tanks as precipitation. For fermentation it is desirable that the water should be slightly alkaline. Hence the effluent from the precipitation tanks must receive a small dose of lime, of urine, or of wood ashes.

As regards the self-purification of rivers more or less polluted with sewage, with the drainage of manured and cultivated lands, and with the waste waters of factories, we find eminent authorities expressing the most conflicting views. Perhaps, indeed, these differences are more seeming than real. Different chemists and microscopists may not attach precisely the same meaning to the terms "pollution" and "purification." Further must be considered the relative proportions of the volumes of sewage, and of the river into which it falls. If a large, pure, and tolerably rapid stream, receives at some one point sewage to the extent of, say, one twentieth of its own volume, we have evidently one set of conditions. But if a small, sluggish stream, receives foul waters all along its course, and that, moreover, in large volumes, the case is essentially different. What is true in the one instance may be utterly untrue in the other.

It cannot be denied that, unless the pollution of a

stream has reached a very great height, there are certain natural agencies which very much reduce the extent of its impurity. Some years ago, an eminent French chemist, Gerardin, proposed to estimate the relative purity of waters by the proportion of free oxygen, which they were found to contain in solution, and which he proposed to determine by the method of Schutzenberger and De Lalande. The details of this method need not occupy us here. Among other waters which he examined were those of the river Vesle, above and below Reims. As this town is the seat of thriving woollen manufactures, the Vesle is polluted in a very similar manner to the Aire at Leeds, with refuse of the most varied kinds, organic and inorganic. The investigator found that at a few miles above the town the water contained a normal proportion of free oxygen. Approaching to Reims, and receiving the waste waters of the woollen mills, it showed a smaller and smaller percentage of free oxygen and a minimum in this respect, with a corresponding maximum of organic pollution was reached a little below the town. Afterwards, as the stream continued its course through an open agricultural country, the proportion of oxygen again increased, and at the distance of about 20 miles below Reims it showed the same percentage of dissolved oxygen as had been observed above the town. At the same time there were observed a series of changes in the organic forms inhabiting the river, which seemed to proceed step by step with the decrease and then with the increase of oxygen dissolved in the water. To these changes I must beg to call special attention elsewhere. Our present concern is merely with the fact that the river Vesle became polluted by the filth of a manufacturing city, and afterwards, from a chemical point of view, at least, returned to its former condition without having been submitted to any artificial process of purification.

We may take another case nearer home. In the year 1868, with a view of throwing light upon a scheme in contemplation for purifying the waste waters of some extensive dye-works, the present writer visited a branch of the Calder and Hebble navigation, which extends from the lower end of the town of Halifax down to the village of Salter Hebble, about 1½ mile off, where it joins the main trunk of the Calder. The descent is so steep that the canal is merely a series of pools separated by locks, and at its origin in the town it is fed not with water, in the ordinary sense of the word, but with sewage from the shallow river Hebble, which receives the refuse, domestic and manufacturing, of the town. Having inspected the canal basin, where the water was very foul and turbid, and emitted an offensive smell, I walked along the towing-path, carefully noting the phenomena manifested. For the first two or three locks there was no material change; no animal life could be detected in the water, and no plants were seen except sewage fungus, which had here and there attached itself to stonework, piles, etc. When about half the length had been traversed water weeds began to make their appearance—at first few and by no means flourishing. They quickly became more numerous, both in individuals and species, as well as more healthy looking. The water exhibited a corresponding change both in colour and smell. The canal, it must be added, was very sparingly used, and its successive pools might be regarded as a series of subsidence tanks. Hence it appears that subsidence and exposure to air alone will bring impure water—in the absence, at least, of certain mineral poisons—to a state which permits of the existence of aquatic plants and of insect life. This point being once reached, a further and considerable improvement is effected by these natural agencies.

Another case of self-purification, effected on a far larger scale, though in a water probably much less polluted than that of the Hebble, as just mentioned, is described by Prof. A. R. Leeds, Ph. D., in his report on the water supply of Wilmington (Wilmington, Delaware: James & Webb). This eminent and experienced chemist goes further than I should venture, since he pronounces that "there is no foundation in fact for the oft-repeated statement that water once polluted by sewage can never again become safe for drinking purposes." He relies on the various natural purifying agencies constantly at work, giving especial prominence to the action of aquatic plants, and to that of the finely-divided earth washed into rivers by the rains, the former influence oxidising—in other words, burning up—and the latter occluding and precipitating the organic impurities. In proof of the efficacy of these natural agencies, he refers to the River Passaic, pure above the town of Paterson, but polluted by the sewage and the manufacturing refuse of that town. He writes:—" The river immediately below the town is black with dye-wares, the fish carried over the great falls are immediately poisoned, and I have seen the foul-smelling and disgusting water covered with their floating carcases." Yet samples of water taken at intervals of a mile apart down the river show a regular decrease in the organic pollution, and an improvement in the quality, until at a point 16 miles below Paterson the river has returned to a condition of purity not much inferior to that which it exhibited above Paterson, and at this point it is used as a water supply for the 300,000 inhabitants of Jersey City and Newark.

Another case of self-purification in a river is reported by Dr. F. Hulwa, in the *Gesundheits Ingenieur*. He has found experimentally that the water of the Oder, after receiving the sewage and other liquid refuse of Breslau, is purified by the oxygen of the atmosphere, and by the action of vegetation. Fourteen kilometres (about 8 miles) below Breslau the sewage matters can neither be detected microscopically nor chemically, and the water of the river was of the same quality as above the town.

An excellent illustrative case is supplied by the Seine below Paris. Immediately below Paris, say at St. Ouen, the water is very filthy; at Epinay and Argenteuil fish reappear. At Beyon water plants grow luxuriantly, whilst at Meulon and Vernon pollution may be said to have disappeared, and the water is actually purer than it is at some distance above Paris. The total nitrogen in the water at Agnières above Paris (not to be confounded with Asnières below the city) is 1·5 gramme per litre; at Clichy and St. Denis 4·0 to 7·0 grammes; at Meulon and Vernon respectively 2·2 and 1·4 gramme.

Further instances of this kind might be brought forward were it at all necessary. But those already given may be fairly held to prove that, under certain favourable conditions, polluted river waters may be purified by natural agencies without any special human intervention. It may be well to fix the degree to which such purification may extend. I have seen no instance where a river when once polluted with sewage has become spontaneously fit for drinking and cookery. But I should be very reluctant to drink the effluent from the most successful precipitation process, passed, for additional security, over a new irrigation field. Yet self-purified rivers may reach such a relative point of purity that they no longer offend the eyes and the nose, and that their proximity has no perceptible effect upon public health. Such a stage is, perhaps, the utmost that can be demanded from rivers, even in uninhabited countries.

But the relative purity in question is reached only under certain special conditions not everywhere met with. The volume of polluting liquid must be small compared with that of the river itself, and, above all, fresh supplies of sewage must not be introduced every couple of miles. These conditions are plainly wanting in the case of English rivers in the manufacturing districts. They are, comparatively speaking, small, their flow is, in the summer, often scanty; and even when they do not pass through towns their banks are garnished with a succession of cotton and woollen factories, chemical works, dye and print works, etc. Thus the natural purifying agencies, whatever their efficiency, have no opportunity to come into play. Further, so much of the matter poured into these streams is of a poisonous character that water weeds, the great oxidising agents, cannot grow. Even the dreaded sewage fungus (*Beggiatoia alba*) may be unable to flourish.

But the great reason why we cannot trust to self-purification in the case of rivers is that it requires prolonged exposure to the air, during which time the water is unfit for almost every imaginable use, domestic or industrial, and the vapours given off are offensive, if not positively dangerous to health. We want a much more speedy system of treatment, and which shall not in the meantime create nuisances as great as, or possibly greater than, those which it is intended to correct.

CHAPTER XIV.

DETECTION OF SEWAGE POLLUTION IN RIVERS AND WELLS.

WHERE the degree of pollution in a stream is excessive —whether caused by town sewage in the ordinary sense of the term, or by industrial waste waters—no special methods of investigation are needed. An ordinary pair of eyes, an ordinary nose, with the will to use them, and to tell *truthfully* what has been observed, is fully sufficient. If the water of a river, viewed in bulk, looks blackish and opaque; if bubbles of gas are seen ascending to the surface and bursting, occasionally, perhaps, bringing up with them lumps of filth, and if in calm weather the smell is decidedly offensive, no further inquiry is needed. As instances may be mentioned the Irwell, Irk, and Medlock at Manchester, the Clyde below Glasgow, the Mersey at Widnes, the Soar at Leicester, the Calder at Brighouse and Mirfield, the Aire at Leeds, and, generally speaking, the rivers below any of our manufacturing towns. But there are many cases where the degree of pollution, though less excessive, is yet enough to render the water unfit for all domestic and for most manufacturing purposes. To judge of these, attention must be specially directed to certain points which may easily escape the attention of the general public. I must, therefore, invite my readers to form themselves into a volunteer "Rivers' Pollution Commission," unpaid, but honest, and not mainly in search of evidence to support pre-conceived notions. Let them then kindly accompany me in an imaginary stroll along the banks of a stream.

We will first take the river as it flows through a rural district. We do not, of course, find here the transparency of the Leven where it issues from Loch Lomond, of the Dudden, or the Derwent. But we see fishes darting through the waters, the whirligig-bettle is spinning merrily round, singly and in little groups; Dytisci plunge to the bottom on our approach, water-boatmen and water-scorpions are searching for prey, whilst brilliant dragon-flies are wheeling on high over the stream. The shallows are lined with arrow-heads, reeds, flags, and the water-iris. In quiet pools the water-lilies lie at anchor, spreading out their broad leaves, and, if we have come at the right season, their beautiful flowers. In short, animal and plant-life, adapted to the water, are to be seen in all their variety and luxuriance.

We dip up a small glass jar of the water and hold it up to the light. We might not, perhaps, select it as a beverage, or like our food to be cooked in it. It receives, undeniably, the drainage from ditches which traverse manured fields. Here and there, too, we may have seen a cottage near the banks, which may have contributed its driblet of animal and vegetable contamination. *May*, we say, because the night soil, the soap-suds, and other impurities from such cottages are more apt to find their way into the garden plot than into the stream. But, in spite of all such possible pollutions, the water looks fairly clean, and gives off no smell, and as we throw it away, we must admit that numbers of our countrymen of all ranks are drinking worse water, did they but know it.

Now let us suppose that the stream, in accordance with that "beneficent arrangement of Providence" which has moved a modern rhetorician to an eloquent outburst, approaches a small town, free from manufactures and unharrassed by "injunctions" concerning river pollution. Here it will be found that every householder uses the stream as a general receptacle for liquid refuse, and too often for solids also. If we now carefully examine the river below the town, where the banks are no longer interfered with by buildings, we shall see that a change has taken place. The water has become distinctly duller, more opaque, less free from colour, and gives out a smell, not yet exactly to be called a stench, but tending in that direction. If we stir up the mud at the bottom with a stick, we shall generally see bubbles of gas arising. Most especially a change comes over the character of the vegetation. Some species will have disappeared altogether; others exist, but they do not flourish.

Here, however, we must point out an error into which not a few authorities have fallen. It has been believed that watercress, growing along the margin of a stream, or in ditches communicating with it, is a proof of the absence of sewage pollution. This assertion at once struck me as being little in harmony with my own observations and those of my friends. Mr. C. Cresswell, Q.C., of Isleworth, a gentleman well-known for his intelligent zeal in the cause of sanitary reform, came upon a ditch which received the entire sewage and household slops from a row of about a dozen cottages. Yet it was filled with the most luxuriant watercress, which, sad to say, appeared to be regularly cut for sale. To decide the matter, I performed a somewhat extensive series of experiments at Aylesbury in the summer and autumn of the year 1878. Earth was placed at the bottom of four small movable tanks, and in each were planted healthy roots of watercress. The tanks were then filled respectively with town sewage, undiluted, and taken directly from the sewer mouth; with similar sewage after treatment with clay, carbon, and aluminium sulphate; with water from a feeder of the river Thame, which flows past the town, receiving the drainage of manured and cultivated lands, and, I believe, sewage from scattered cottages along the upper part of its course, and with the excellent drinking water supplied by the Chiltern Hills Company. The loss by evaporation or leakage in each tank was made up by the regular addition of the same kind of water as had been taken at first. The result did not admit of a moment's doubt; the watercress planted in sewage not merely lived and grew, but far surpassed the other three lots in luxuriance and vigour, and continued so to do until the experiment was broken off by the frosts in the beginning of winter.

Hence it may surely be concluded that the presence of watercress in any stream or pond affords no proof of the purity of the water. It must be understood, however, that the sewage made use of, though unusually rich in fæcal matters, blood from slaughter-houses, and other animal pollutions, was perfectly free from manufacturing refuse. I have never come upon or heard of any case of watercress found growing in a stream which receives the waste waters of chemical works, dye works, woollen factories, etc.; but as the sewage in question contained very much less free oxygen in solution than the Chiltern Hills' water, or even the river water, it is plain that the growth of the higher vegetation, *i.e.*, green plants as distinguished from grey fungi, does not follow, step by step, the rising or falling proportion of such oxygen in the water. We may, indeed, ask why should it? It can scarcely be needful to remind the reader that all green plants—in other words, all plants which develop chlorophyll in their tissues—give off oxygen on exposure to the sun, or even to diffused daylight. Hence water plants are not so much the effect as the cause—or at least one great cause—of the presence of free oxygen in water. Their absence in highly polluted streams is due not so much to a deficiency of oxygen, as to some positive injurious agency, whether acids, alkalies, or metallic compounds. In the absence of such plant-destroyers, green vegetation is very efficient in freeing water from excrementitious pollution. There are, of course here, certain limits ; solutions of putrescent animal matter may be too concentrated to admit of vegetable life, just as liquid manures may be applied in too strong a dose. This is a point rarely reached in any stream, and when green plants are absent, we may generally seek the cause in the waste waters of some manufacturing establishment. Town sewage, when treated by any process which leaves in it a large proportion of any compound of lime, is also unfriendly to aquatic vegetation. The sanitary authorities of a large manufacturing town were advised by the late Mr. Smee to plant *Elodea Canadensis* in the outer tank of their sewage works, and to attempt the growth of reeds and sedges around the margin. The advice was in itself excellent, inasmuch as such vegetation would have completed the purification of the sewage. But, as the process adopted involved the daily consumption of about fifteen tons of lime, we can scarcely wonder if the plants failed to survive. As far as the writer has observed, natural streams of very hard water are not rich in aquatic vegetation.

We have now to turn to the animal world, and ask whether the presence or absence of fishes can be said, in general terms, to be solely or mainly dependent upon the greater or smaller proportion of free oxygen in the water? Or to put the same thing in a slightly modified form, upon the smaller or larger proportion of decomposing organic matter in the stream, since we have already noted that where such organic matter is most abundant, there free oxygen must exist in the smallest proportion or be entirely wanting. That free oxygen, or, as it is commonly called, a "good aëration" of the water, is necessary for fishes must be known to even the most careless proprietor of an aquarium. We may, therefore, safely say that if free oxygen is wanting, fishes will be wanting likewise. But can we draw the converse inference, that if fish are wanting oxygen must be deficient? By no means. There are various substances which occasionally find their way into rivers, and prove very widely fatal to fish, but which are not calculated to effect any decrease in the proportion of dissolved oxygen. We may take an instance given on very good authority, and briefly recorded in the *Journal of Science* for 1880 (p. 213). Dr. Auerbach during an entire summer observed certain water-beetles—from his description evidently *Gyrinus natator*—living in tanks full of a saturated solution of sodium sulphate (Glauber's salt). When alarmed, these little beetles dived down, and hid themselves among the crystals that were forming, just as they would do among the plants in a pool. But a little of this liquid, thus harmless to certain insects, happened to escape from the tanks by leakage, and found its way along a ditch into a river at some distance, where it proved fatal to a multitude of fish. It can certainly not be contended that a small quantity of a solution of sodium sulphate added to the water of a river would either appropriate or expel the free oxygen, yet we see that it turned the scale between life and death.

Cream of lime thrown into a stream will, as it is universally known, seize upon carbonic acid, leaving any oxygen it may encounter unaffected.

The presence of fish is certainly no proof of the absence of fæcal matter. At Kingston, just where the sewage of the town entered the Thames, in the spring of 1875—it may possibly have since been diverted—I have seen fish darting about in numbers, and I learnt that the sewer mouth was a favourite spot for anglers. But where the volume of sewage is small as compared with the volume of the river into which it flows, such pollution by no means necessarily involves a deficiency of free oxygen. In recent sewage there are also various matters which certain fishes will eat greedily, though with doubtful benefit to their health. At least it is asserted that fish caught in polluted waters enter into decomposition with exceptional rapidity, a probable indication that their entire system is in a bad condition.

Among amphibious creatures frogs are found in pure waters, and in those but slightly polluted, and the same may be said of newts. If either excrementitious or manufacturing refuse is introduced in appreciable quantity, they withdraw.

The presence of aquatic insects is not a character from which any definite conclusion can be drawn, though where industrial waste waters are to be found in quantity they are generally wanting. Thus I have never seen either larvæ or adult insects in the Bridgewater Canal, or the Irwell below Manchester; in the notorious Sanky Brook, in the Aire below Leeds, the Kelvin Water, or similarly polluted streams. But in rivers fouled with putrescent vegetable refuse or fæcal matter they often abound. On the contrary, in the very purest water they are necessarily absent, as finding there no food. It would be interesting to find what is the minimum of impurity at which the larvæ of gnats and bloodworms (*Cheironomus plumosus*) are able to exist, and what is, if any, their maximum limit.

These larvæ are frequently found in water-butts and cisterns, which have no other source of contamination but the organic matter of dust. Still, I would suggest that no water, in which these larvæ are present, should ever be used for domestic purposes, since their excretions, as far as the writer has been able to observe, set up in organic matter decomposition of an exceedingly offensive and probably dangerous type. This character of their juices may possibly explain the irritating and, in some cases, even dangerous effect of the bite of sand-flies, mosquitoes, and pollution-fed diptera in general.

Water beetles, such as *Acilius sulcatus*, *Colymbetes* of various species, etc., and also certain Hemiptera, may be found in water, which, from the absence of known sources of pollution, and from chemical and microscopical examination, may be pronounced potable. But they are also met with in pools fed by the surface drainage of manured fields and pasture lands. As a rule, it may be said that animals which prey upon living animals or upon growing plants are not, in themselves, a bad symptom. All such as consume comminuted or putrescent matters, animal or vegetable, must be considered questionable, as proving the presence of the matter upon which they subsist.

But we must now return to our river, once fairly pure, but now polluted by its wanderings among the dwellings of man. Let us suppose the contaminations increase to such a degree that the green water plants become perceptibly less plentiful and less flourishing. If we now look carefully into the water at the margin of the stream and examine piles, piers of bridges, the roots of trees projecting under water, we shall see a growth, which is considered most characteristic of polluted waters. This is so-called sewage fungus (*Beggiatoa alba*), a plant which is much talked about, but which many redoubted sanitary reformers and patentees of processes for the purification of sewage appear never to have seen. They often seem to suppose that it must have something of the appearance of a mushroom or toadstool. This is a complete error; in form and colour it is not unlike a bundle of tow, with the fibres running parallel to each other and ending in loose tufts. Suppose such a bundle fixed at one end to a stick, or a stone, or to the earth at some little depth under water, and swaying to and fro in the current, and you have a fair resemblance of sewage fungus. The chief difference is that this unholy and unlovely plant has a greater specific gravity than hemp and flax, and tends to sink rather than rise if not kept in a horizontal position by the current. The colour, too, is modified by the particles of dirt which get entangled in the fibres. What concerns us is neither the structure, nor the affinities, nor the chemical composition of this fungus, but the conditions under which it exists, and its consequent value as a sign of water pollution.

In the first place, contrary to the common notion, it is not peculiar to sewage. It has been found in the North of England in streams quite free from animal pollution, but which receive the drainage from the heaps of vat-waste—the unpleasant residuary product of the alkali manufacture on Leblanc's principle. It is found in the escape waters of the medicinal sulphur springs of the Western Alps and the Pyrenees. In short, what it indicates is not necessarily animal pollution, but sulphur. Without sulphur it cannot live; and where it thrives, sulphur in some form is certainly present.

It is almost, without exception, peculiar to running water. Only once have I seen it formed under my eyes in a jar of impure water. I have never seen it at the sides of any pond or reservoir, whether of pure or of polluted water, nor at the bottom of such pools, if emptied. If planted in an aquarium for experimental purposes, it hangs straight downwards from the stick or root to which it has been found attached, and whether kept in the light or in the dark, it shows no disposition to spread. Moving water seems, therefore, essential to its growth. If preserved in the dark it undergoes no change for months, and seems unaffected by the most powerful chemical agents, except in enormous proportions—certainly by no quantity which could be added to the waters of a river. Chromic acid is not reduced by it, even on prolonged contact. If, however, sewage fungus is placed in still water and exposed to a strong light, green confervæ fix themselves upon it, overspread it, and seem gradually to effect its destruction. Similar cases may be observed in shallow trenches, in which partially purified sewage is flowing ; but where water is still strongly charged with animal matter the fungus appears to hold its own, especially if the depth and turbidity of the stream interfere with the free action of light. It need scarcely be said that sewage fungus is never found in pure mountain streams, and but very rarely in the brooks and ditches of rural districts which receive the drainage of cultivated lands, except where they are connected with some sewer. But if portions of the plant are swept down out of a sewer—*e.g.*, by a violent storm of rain—they are able to live in running water where the amount of pollution is exceedingly small; smaller, indeed, than the proportion specified as to be tolerated in the far-famed " recommendations " of the late " Rivers' Pollution Commission." If it be asked how this is ascertained, I reply that I know a small stream, which, down to a certain point, is absolutely free from sewage fungus, as I have satisfied myself by careful and repeated examination. At that point it receives a stream of sewage effluent about the one-sixth part of its own volume, and purified to such a degree that one of the highest authorities on water analysis, after repeatedly examining samples taken at haphazard, has pronounced it to fall *well within the limits* of the recommendations just referred to. Yet the stream, after receiving this infinitesimal dose of fæcal matter, displays here and there a tuft of sewage fungus along with a most luxuriant growth of green water plants. It must further be noticed that, on following the stream downwards for a few hundred yards, the fungus disappears, its pabulum having, without doubt, been burnt up by the oxygen evolved by the higher plants.

But whilst water may thus be too pure to supply the sulphur and the other nourishment required by sewage fungus, it may also be found, if not too impure, at least, not to possess the right kind of impurity. I have made frequent and most minute examination of the sewage of Leeds, and of the results of the various processes which have been adopted for its purification. But I have never seen a trace of sewage fungus, either brought down the sewers from the town or floating in the tanks, or in the outflow channel, or where the latter vents into the filthy river Aire. This absence was always distinct, whatever was the nature and the success of the process adopted.

Again, the London sewage at the southern outfall has never, in so far as I have had the opportunity of observing, contained a trace of sewage fungus.

The same may be said of the sewage of Paris, which I have studied daily for some weeks at Gennevilliers. On the other hand, a ditch at Wimbledon, which received the sewage of the district, some of it treated and some of it untreated, contained in the autumn of 1875, some beautifully characteristic specimens. At Aylesbury, some years ago, it was swept down the sewers from the town after heavy rains, and arrived at the sewage works in quantity. For the last few years it has almost totally disappeared. At Hertford, it has been seen in great perfection in the outflow channel.

Hence we may probably conclude that sewage fungus grows by preference in the rich sewage of residential towns rather than in the waste waters, unless sulphuretted, of the great manufacturing centres, or in the very dilute sewage of London or Paris. When once fairly established, it is, however, able to live on in relatively pure water. It need scarcely be said that this fungus, containing no chlorophyll, gives off no oxygen on exposure to light, and, consequently, does not contribute, as does green vegetation, to the purification of the water which it inhabits. Of course it withdraws a certain quantity of organic matter from the water so as to form its own tissues, but this, on its ultimate decomposition will be returned to the stream.

It does not appear to form the food of any animal, certainly not of fishes, insects, the higher crustaceans—such as *Astacus fluviatilis*—and mollusca. Infusoria swarm among its fibres, probably as a place of shelter. Nor can I learn that it has ever been tried as an article of human diet. It might not here be impertinent to express the wish that a certain person, who, some years ago, wrote to the papers suggesting rats (*sic !*)—foci of Trichinae—as food for the destitute, would kindly make an experiment with sewage fungus *in corpore suo vilissimo*. It is certainly nitrogenous, possibly nutritious and delicate, and just as possible poisonous.

Summing up the foregoing, we may fairly conclude that neither vegetable nor animal life varies in any simple relation with the quantity of free oxygen found in the stream ; that all grey or whitish plant growths, when existing alone indicate great pollution ; but where they occur along with green plants, the impurities may be very trifling. Yet, even vegetation of a relatively higher grade, such as the watercress, gives no positive proof that a stream is potable and wholesome. Perhaps the worst sign of all is the total absence of all vegetation, except its existence is rendered impossible by the rapidity of the current, or by the nature of the banks. Lastly, it is impossible to argue from ordinary town sewage, to industrial waste waters, or to a mixture of the two. The influence of the former upon animal and vegetable life is indeed distinct from that of the latter.

I have purposely avoided any attempt at giving instructions for the chemical or the microscopic examination of waters. The due performance of such operations can be learnt only by actual practice in the laboratory. Without such training, the experimentalist will find his results utterly untrustworthy, and misleading. It sometimes even happens that experts of undoubted skill and wide experience, arrive at results which are far from accordant. I have heard of three identical samples of a water being sent to three analytical chemists who have made such investigations their speciality. One of the three authorities declared the water quite unfit for drinking, but fit for washing. The second pronounced it admissible for drinking, but not suited for washing; whilst the third expert, taking higher ground, condemned it for both purposes.

Considerable weight has of late been laid on a microscopic determination of the number of living organisms present in water as a clue to its quality. Thus, Professor Koch, in a recent publication, declares that "an abundance of microbia in a water is proof that it must have received an influx of liquids rich in decomposing matter, and containing, among many minute organisms, which may be innocent, possibly others which may be pathogenetic, *i.e.*, which may constitute the infection of diseases." He continues :—"As far as present knowledge extends, the number of micro-organisms in good waters varies from 10 to 150 per cubic centimetre. If the number exceed this limit, the water must be suspected as contaminated. If it reaches 1,000, the water should be at once condemned, especially in times of cholera epidemics."

Dr. Link, on the contrary, informs us in the *Archiv der Pharmacie*, and in the *Chemiker Zeitung*, that he has examined a number of the well waters of Danzig, both chemically and microscopically, the results being not in accord with the opinion of Koch as above quoted. On the contrary it became manifest that there prevailed no regular relation between the results of the chemical analysis and the number of bacteria as recognised by the microscope. Many good well waters not directly or indirectly exposed to animal contaminations contained great numbers of micro-organisms. On the contrary, others, which had been found bad on chemical examination, and had been undoubtedly polluted with sewage, contained only very insignificant numbers of bacteria undergoing development. We also consider that the great majority, if not the totality, of bacteria in a well water are probably of a harmless character, and that even if a water is contaminated with pathogenic germs, these will not generally find in well waters the conditions necessary for their multiplication, especially in default of a temperature bordering on that of the human body, and a due concentration of nutritive matter. Hence it follows that the number of the micro-organisms present does not warrant us in pronouncing on the character of the water, and may lead us to conclusions in opposition to the results of chemical analysis. The attempt to put forward microscopic examination as a decisive means of ascertaining the quality of a water, is for the present devoid of an adequate foundation.

To this criticism it may, however, be replied that chemical analysis of waters in like manner fails to show whether their carbon and nitrogen are in the state of harmless or pernicious compounds.

One very simple test which may be applied to any apparently pure water, is to put a few ounces of the sample in a perfectly clean bottle, close it with a glass stopper, or with a new clean cork, and let it stand for some days at a temperature of 60° to 70° Fahrenheit. If on unstoppering the bottle it is found to give off an unpleasant smell the water must be condemned.

This test applies as much to the water of wells and springs as to that of streams.

In wells we are deprived of the help of certain signs which in a stream indicate impurity. There is, or should be, no visible life, vegetable or animal. The water must be rejected for a domestic supply if it contain the larvæ of gnats, or the so-called blood worms. A very bad sign is, if the water after having been undisturbed for some time, *e.g.*, early in the morning, displays an oily scum on its surface. Minute bubbles of gases, adhering to the sides of a glass in which the water has been allowed to stand, are always to be regarded with suspicion, since they may be due to the putrefaction of organic matter. Some of the most deadly waters are the most bright and sparkling. Wells in porous soils such as chalks, sands, and gravels, are unsafe if there are any cesspools in the neighbourhood. There is at present no legal remedy for the pollution of a well by impurities which leak into it *underground.* I am sorry to say that the Bill brought into Parliament during the session of 1885 by Earl Percy and others, did *not* propose to remedy this capital defect.

This is, perhaps, the place to touch upon proposed regulations as to the quality of the waters which should be allowed to pass into a river. This is a subject which has been abundantly discussed. Every man and woman in the kingdom, almost every child, must have heard of the "Recommendations" of the late Royal Rivers' Pollution Commission which have been so persistently obtruded upon the public for about sixteen years, and which, if last year's Sewage Bill had unfortunately been passed, would have acquired legal sanction.

For convenience sake I must here reproduce them. The Commissioners proposed to exclude :

a. Any liquid containing, in *suspension*, more than three parts by weight of dry mineral matter, and one part by weight of dry organic matter, in 100,000 parts by weight of the liquid.

b. Any liquid containing, *in solution*, more than two parts by weight of organic carbon, and 0·3 part by weight of organic nitrogen, in 100,000 parts by weight of the liquid.

c. Any liquid which shall exhibit by daylight a distinct colour, when a solution of it, one inch deep, is placed in a white porcelain or earthenware vessel.

d. Any liquid which contains, *in solution*, in 100,000 parts by weight, more than two parts by weight of any metal except calcium, magnesium, potassium, and sodium.

e. Any liquid which in 100,000 parts by weight contains, *whether in solution or suspension*, in chemical combination or otherwise, more than 0·05 parts by weight of metallic arsenic.

f. Any liquid which after acidification with sulphuric acid, contains in 100,000 parts by weight, more than one part by weight of free chlorine.

g. Any liquid which contains in 100,000 parts by weight,

more than one part by weight of sulphur, in the condition either of sulphuretted hydrogen or of a soluble sulphuret.

h. Any liquid possessing an acidity greater than that which is produced by adding two parts by weight of real muriatic acid to 1,000 parts by weight of distilled water.

i. Any liquid possessing an alkalimity greater than that produced by adding one part of dry caustic soda to 1,000 parts of distilled water.

To these recommendations a most judicious addendum has been proposed, viz., the total exclusion of petroleum and gas tar products, and the refuse of gas-works in general, and of that of india-rubber works.

I must now beg my readers not to be dazzled by the scientific reputation of the late Commissioners, or, at least of one of them—but to take these recommendations on their own merits, and to weigh them calmly and impartially.

It will at once strike the critic that these recommendations ignore the volume and the condition of the river, and attend solely to the quality of the waters poured into it from manufactories, town sewers, etc. without any regard to their quantity.

Now, suppose a number of establishments emitting abundance of water just within the proposed standards into a small river; will it not speedily be brought into a condition like that of the Kelvin Water, the Aire, or the Irwell? But again, suppose a manufacturer on the banks of a river wishes to get rid daily of 100,000 gallons of a liquid containing an alkalinity equal to $2\frac{1}{2}$ "parts by weight of dry caustic soda in 1,000 parts of distilled water," and therefore exceeding the standard. What can he do to extricate himself from the difficulty? He simply pumps up 100,000 gallons of water from the river, mixes it with his waste water, and lets the whole 200,000 gallons flow into the river triumphantly! On analysis it will be found within the standard, containing alkalinity equal only to 1¼ part of dry caustic soda to 1,000 parts of water. He has therefore, obeyed the law, and yet he has put into the river, to a grain, the same quantity of matter as if he had run his 100,000 gallons directly into the river. Indeed, the purer the river, the easier do such evasions become.

I therefore submit that all "standards" based on the principle of so or so many grains or fractions of a grain being permissible in a gallon, or in 100,000 gallons of water, and anything beyond that limit being contraband, are fundamentally and essentially absurd, as simply courting evasion.

If we come to details, the recommendations do not improve. They are not so much stringent, as grossly inconsistent. Nor, as far as I have been able to learn, have the Commissioners ever attempted to justify the precise limits which they laid down in each particular case. The exclusion of free chlorine and of sulphuretted hydrogen, and the soluble sulphides, might have rationally been carried somewhat further. Nor can we object that the disposal of arsenic is rigidly restricted. But why are the neutral salts of potash, soda, lime, and magnesia to be admitted in any proportion? These salts in certain quantities—those especially of magnesia —are injurious if habitually introduced into the animal economy. And why are all the other metals and their solutions placed upon one dead level—2 parts by weight in 100,000? Some of these, such as aluminium, strontium, and iron, are harmless to animal life in even larger proportions. On the other hand, chromium, zinc, and lead, are formidable poisons, and in manufacturing districts are not unlikely to find their way into rivers. Surely 1·4 grains of metallic lead per gallon in a water supply would be amply enough to occasion lead poisoning.

With organic pollutions there is the same inconsistency. One part of suspended organic matter is allowed per 100,000, but only 0·3 part of "organic nitrogen" in solution. Yet "organic carbon" in solution is allowed up to 2 parts per 100,000. How is this? "Organic nitrogen" in solution is, therefore, in the judgment of the late Commissioners, more dangerous than "organic matter" in suspension. Yet "organic carbon" in solution is *less* dangerous than "organic matter" in suspension. Dissolve organic matter, and it becomes less dangerous as regards its carbon, but more dangerous as touching its nitrogen! Surely this is a hard saying.

Again, all sorts of organic matter are not equally prone to decomposition, or equally dangerous when decomposing. The suspended organic matter in a water might consist of stearic acid or of woody fibre, or it might be composed of solidified albumen or of the fibrous matter of blood. Would the danger to public health in these two sets of cases be equal? Yet of such differences, and of numbers of others, the Recommendations take no cognisance.

It is, therefore, I submit, the duty of the public to dismiss them as impracticable, unpractical, and even dangerous, and to propose some simpler standards, less elaborate, and turning less on disputable analyses.

Not a few chemists of merit incline to something like the following scale :—

(*a*) The effluent or any water turned into a river shall be clear and colourless if examined in a cylindrical pint bottle of white glass.

(*b*) It must not be alkaline to test-paper.

Alkalinity is more dangerous in waters than acidity, as it favours putrefaction.

(*c*) If to a pint of the water there is added 1 grain of sulphate of alumina or of alum, previously dissolved in 100 grains of water, there shall not be any perceptible turbidity produced within half-an-hour.

(*d*) If a pint bottle is half-filled with the water in question and well shaken up after standing for 10 minutes, no foam shall appear.

CHAPTER XV.

RECOGNITION OF THE DEGREE OF PURIFICATION REACHED IN SEWAGE TANKS.

IT is often important to judge in how far the effluent water contained in a sewage tank or in the outflow channels has been purified. The chemical tests proposed by Professor Frankland, Professor Wanklyn, Dr. Tidy, and others, whatever may be their respective values, do not permit of an immediate decision on the spot. We have, therefore, to take recourse to a variety of indications, which require careful observation.

In the first place must come the colour of the water. The purer water is, the more its colour inclines, when seen in large masses, to a transparent blue. If, on the other hand, it is rich in dissolved impurities, it verges upon a brown. This distinction has been very clearly brought out in the joint investigation of the London water supplies, undertaken by Mr. Crookes and Drs. Odling and Tidy.

The blue tint of a pure water is something peculiar; it is not whitish-blue or milky, but borders slightly upon a green, and is combined with a very high degree of transparency. To observe the colour, it is well to make the complete circuit of the tank, so as to view the water in all directions with reference to the light. If there are brick buildings near at hand, dark clouds overhead, or if the wind is very strong, an erroneous impression may be made.

In addition to examining the water in bulk as it stands or flows in the tanks, a portion should be dipped up in a clean glass and inspected both by reflected and transmitted light. The presence or absence of scum, of floating particles, and of any colour or opacity, should be carefully noted.

The next point is the odour. Of all man's senses, that of smell is generally the vaguest in its impressions. Many people smell, or profess to smell, not the odours which actually exist, but those which they think might, could, would, or should, be perceptible. It has been said that a scent which would pass unnoticed in a drawing-room, will at once excite hostile comments at sewage works. The present writer once met with an instance of this kind. A very strong east wind was blowing, and carried to and over a certain sewage works the smell of a factory locally known as "Honey-pots' Hall," where animal offal of various kinds was treated for grease, glue, manure, etc. A visitor who had no faith in precipitation processes, and who maintained, *con strepitu*, that "even the best of them do not purify but only clarify," complained that the smell came from the effluent! He was prevailed on to walk on until he was more than a hundred yards to the windward of the tanks and the effluent channel, and was asked whether the stench could possibly travel such a distance in the teeth of a strong wind? His reply was characteristic: "I still adhere to my opinion." Such a man was by nature qualified for a "Royal Commissioner," who can never admit that either himself or any of his predecessors can have been mistaken.

If any person wishes to form an honest opinion as to the presence or absence of smell from effluent, I should recommend him to walk carefully round the tanks and along the channels. The fairest judgment can be formed when the mouth of the sewer is to the leeward of the observer, so that he may not ascribe to the effluent the offensive smell of the raw sewage. It is sometimes the practice to bottle samples of the effluent, stopper them and set them aside for future examination. In such cases it is well to take, on the same day, accompanying samples of water from some ordinary river or pool which does not receive excremental pollution, and compare the smell of the two! The proportion of fair average drinking waters, which, when thus preserved for a time in closed bottles, do not give off a disagreeable smell is smaller than might be either desired or expected.

Another point to be observed is the vegetation which may be found clinging to the wood or stone work of the tanks and the outfall channels; if these are fringed with sewage fungus it is a sign that the purification of the sewage still leaves something to be desired. But if stone work just about the level of the water or below it in shallow places is covered with green confervæ, the process in action may be considered good. The confervæ will carry on the good work by means of the nascent oxygen, which they give off under the action of light, and which burns up, or in technical language, oxidises the remaining organic impurities. Hence persons in charge of sewage works should never sweep away such green growths. By so doing they merely destroy a useful fellow servant.

It is always desirable to watch the behaviour of the

water where it flows over a lip, down a steep incline, or passes any obstacle which causes the formation of bubbles or froth. If such bubbles break immediately, the effluent may be regarded as well freed from dissolved organic impurities, or, in other words, as being not merely "clarified but purified." But if the froth is persistent and may be traced a considerable distance down the channel, much soluble organic matter is still present. This test, however, gives very little clue to the nature of the impurity. Thus a spent lye from soap works, containing glycerine, favours frothing as much as does gelatine, mucus, albumen, or other putrescible nitrogenous matter.

If practicable, it is well to follow the outfall channel down to the stream into which it discharges, examining both the appearances along its course and its effect, if any, upon the water of the river. The points to be more particularly considered are the occurrence of a secondary precipitation in the channel. This, if it takes place, is a sign that the tanks are too small, or the incline too steep, so that the water after treatment has not sufficient time to settle. A defect of this nature is no evidence against the process in itself, though it affords proof of defective arrangements.

But if a fresh precipitation takes place where the effluent water mingles with that of the river, we may generally infer that some precipitating agent used for the sewage has been added in excess, so that it goes on precipitating the suspended and dissolved solids in the stream, which action will not take place if the water of the river is very pure, containing practically no solids. Nor will it be noticeable in the case of a very foul river containing much manufacturing refuse. But in the streams of lowland rural districts it may be very striking, and may give rise to complaints that the effluent water from the sewage works is polluting the river. I have seen a case of this kind, where the sewage of a large suburban village and district was first treated with lime in precipitation tanks of the ordinary structure, and the effluent was then used for irrigation on a large plot of land. Yet the outflowing water on mingling with that of the little river which received it, occasioned a decided increase of turbidity, and there was an outcry accordingly. But the river was full of vegetable pollution, the drainage of extensive woodlands, and of a large stretch of fields, pastures, and market gardens, and the effluent from the sewage works merely rendered these impurities more conspicuous.

Another point of no small importance is the action of the effluent upon fish. Do they, if the level allows, ascend the outfall channel without suffering any apparent inconvenience? Are they to be found in the river *below*, as well as *above*, the point of outfall, or do they carefully avoid coming in contact with the effluent? These are questions of great importance. If the effluent is fatal to fish, or if it at least is evidently an annoyance to them, we may conclude that the sewage is insufficiently purified, or that some improper agent is used in its purification.

A few hints may here be given about sampling effluents for analysis, or for preservation. I write here mainly for the guidance of officials left in charge of sewage works, that they may know what tricks may be attempted. It is commonly said "that any stick is good enough to beat a dog with," and, on a similar principle, any unfair, dishonourable stratagem is held legitimate to bring discredit on a process for the chemical treatment of sewage.

I must first speak of the construction of the works Everything should be so arranged that it should be impossible to take a sample surreptitiously. The manager should always be able to state upon oath, if required, whether any given person did on some specified day receive a sample or not. I have known samples of water fraudulently concocted and then analysed with all due solemnity, whilst all the while no such water was at the time running in the stream from which they were said to have been taken. A certain very astute contractor for treating sewage, always takes good care that the effluents on leaving the enclosure of the works, flow in a covered channel, and issue into the river *below the surface*. Thus the surreptitious procuring of a sample is rendered impossible.

Another needful precaution is to admit no visitors, and allow no sampling at times of flood, unless the tanks and other arrangements are fully capable of dealing with the unusual quantity of storm-water. To pay visits of inspection at such times, and omit in subsequent reports, all mention of the abnormal state of things, is a device not quite strange to the official mind. We believe a person once paid a visit of this kind to the whilome Leamington Sewage Works, and sat in a cab, which was not quite lifted off the ground by the flood, watching the working of the process through the window. No visitors should be allowed who have not the opportunity to examine closely and carefully, and at the same time have not the candour to confess under what anomalous circumstances their inspection has been made.

Another necessary precaution relates to sampling. If any visitor asks for a sample, it should be given him solely on the condition that he takes and seals up at the same time a check sample, to be left at the works for analysis by some independent chemist. Without this precaution he may, for instance, add to his sample a little urine or blood, or a culture solution, and still represent it as a normal sample.

The bottles used must be perfectly clean, by preference new, and fitted with glass stoppers. They should have been well washed out with plenty of a fairly good drinking water, and be then rinsed out twice or thrice with the effluent to be sampled. No one should be permitted to take a sample in a bottle which has contained wine, beer, or any organic liquid whatever.

It must not be thought that these precautions are dictated by an imputative jealousy; they are the fruits of experience. I have seen a man come to a sewage works for a sample of the effluent, bringing with him a bottle most palpably dirty. One of the workmen shattered it with a crow-bar, and on examining some of the fragments I found them thickly coated within with organic matter. As the man refused a sample in a clean bottle, I can only conclude that his dirty bottle must have been specially selected or prepared, in order to admit of a false and condemnatory analysis. This is by no means the only case where bottles brought by strangers have been found to be exceptionally dirty.

The act of taking the samples should be done in presence of both parties. If bottles are filled from a channel lined with stone, brick, or concrete, there is no possibility of any jugglery. But I have known instances where it was desired to take a sample at the point of junction where the effluent was discharged into a dirty brook. In two such cases, the person who had come for the sample showed an evident desire to scrape up a portion of the dirt at the bottom of the stream, which would have seriously swelled the sum of "total matter in suspension,"—the more so as prior to the adoption of the process of treatment a large portion of the sewage of the town had been discharged at the very spot in its raw condition.

CHAPTER XVI.

PRECIPITATION MUD.

WITH the production of an effluent, clear and pure as practicable, the difficulties of sewage treatment are not exhausted. The water and the deposit have to be separated from each other, so that, according to a favourite phrase, "the water may go to the river, and the manure to the land." This must be done frequently. If not, the tanks begin to choke, the sediment ferments and rises up, spoiling the water, and the result is one to gladden the heart, and for once justify the representations of ex-Royal Rivers' Pollution Commissioners. Further, the effluent must be drawn off very quietly, without disturbing the mud. Where the slope of the ground allows, this is effected by gravitation. The tank to be emptied is shut out of connection with the remainder, and allowed to rest for a few hours, so as to ensure perfect settlement, and the water is then run off as nearly as possible to the level of the deposit. The pipe through which this is effected ends in several feet of flexible hose (more or less, according to the depth of the tank), and is capped with a funnel *opening upwards*. This funnel is at first placed near the surface of the liquid, and as the process goes on it is gradually and carefully lowered by means of ropes or chains, until the mouth of the funnel, still kept upwards, is only just above the level of the mud. Where there is not a sufficient fall, the water is drawn off by means of a pump, but the flexible hose and funnel are still necessary.

When the effluent water is thus disposed of, the mud is run off, or pumped off, into suitable receptacles for further treatment and conversion into a portable manure. The mud is a thin paste, containing, on an average, 90 per cent. of water. If a good process for precipitation has been used it is, practically speaking, inodorous; it does not, when separated from the water, pass readily into fermentation, and it has but little attraction for flies.

It may be dried by a variety of methods. Where there is a sandy or otherwise porous soil, and where the climate is dry, the mud may be run into a stank, *i.e.*, a large, shallow reservoir, with an earth bottom and sides. Here it forms a layer about 6 to 9 inches in depth. The moisture is given off, partly by evaporation, but chiefly by absorption into the soil below. Under favourable circumstances it may, in this manner, be reduced to 20 per cent. Such a result cannot always be reached in Britain, at least in any ordinary season, and as the process requires considerable space, a long time and much labour, it cannot be generally recommended. The mud-stanks, further, though free from nuisance, if a proper system of treatment has been employed, have an unsightly appearance.

The best method of dealing with the mud is, generally speaking, to run it into a filter-press, such as that of Needham and Kite, of Johnstone, or of Manlove,

174 *SEWAGE TREATMENT.*

Alliatt & Co. The accompanying figures, 2 and 3, represent the filter-presses of the last-mentioned firm.

Here the moisture present is reduced down to 50–40 per cent. according to the pressure put on and the time allowed. The mud comes out on opening the presses

Fig. 2.

Fig. 3.

in flat cakes, circular or rectangular, according to the make of the press, sufficiently coherent for handling.

At many sewage works it is, strange to say, customary to add some substance to the mud before running it into the presses, for the purpose of facilitating the drying process.

Among the materials so used are—

a. Finely-ground basic slag, coprolite, apatite, or other phosphatic mineral. This substance has, of course, an intrinsic manurial value and cannot be supposed to deteriorate the properties of the sewage mud. But it has little solidifying power, and scarcely facilitates drying. When finely ground it is also somewhat expensive, and cannot well, in the resulting manure, be sold for more than it has cost, so that the whole transaction is like giving change for a shilling.

b. Gypsum, or plaster of Paris. Everyone knows that this mineral, at least in the burnt state, readily takes up water, and solidifies it, so that if a sufficient quantity were added to the mud, neither pressing nor any other drying process would be needed. But in contact with carbonaceous matter, gypsum is decomposed with a plentiful escape of sulphuretted hydrogen. This, if occurring at all on a large scale in the neighbourhood of human habitations, may rightly be viewed as a public nuisance, and in any case, it will greatly endanger the health of the workmen. Besides, gypsum may be safely pronounced to have no manurial value, our ablest agricultural chemists regarding it as a "mere diluent."

c. Carbonate of lime, in the state of chalk, or other convenient form.

Here we have certainly no production of a nuisance, as in the case of gypsum. But we have, firstly, a direct deterioration of the manure by the loss of one of its most valuable constituents. The nitrogen existing in the state of ammoniacal salts is liable to be driven off. Mr. T. Brown, writing recently in the *Chemical News*, on "The Failure of Sulphate of Ammonia in Manuring Experiments," points out that ammonia is volatilised even by chalky soils. Of course, if quicklime, or slaked lime, is used, the loss is greater and more rapid.

But even if we suppose that no ammonia were directly expelled by these additions, we must not forget how greatly the value of sewage mud is diminished by the addition of any worthless material for the purpose of drying up. Let us suppose a mud containing 90 per cent. of water, and consequently, only 10 per cent. of solids, more or less manurial. If we add to such a mud 10 per cent. of chalk, we divide the nitrogen present, the phosphoric acid, etc. in a given weight, by 2. If we use 20 per cent. of chalk or other carbonate of lime—proportions which, I believe, are not merely reached, but actually exceeded in practice—we reduce the nitrogen and the phosphates present to one-third their original percentage. Can we then wonder if sewage manures thus treated, are pronounced not worth the cost of carriage, and of the labour involved in applying them to the soil?

Gypsum, if burnt, at least, behaves still worse, since it seizes on and retains one-fifth its weight of the moisture present in the mud. Thus, if we add 20 per cent. of gypsum to sewage mud, we shall have 24 per cent. of rubbish.

Additions of lime, chalk or gypsum, to sewage mud, are, therefore, a complete mistake, as well as being quite needless.

The greatest drawback in the use of the filter press is the exuding water. This is in quantity no trifle. One hundred tons of mud contain 20,000 gallons of water, of which about 9,000 are squeezed out by the press. The disposal of this liquid is a matter of some difficulty. In some places it is passed back into the sewage or into the tanks, and is treated over again. This is by no

DRYING CYLINDER.

To face page 177.

means judicious; the press liquor is far harder to treat than fresh sewage, requiring an extra dose of chemicals, and even then, giving but a doubtful result. Press liquor, in fact, behaves very much like stale, putrid sewage. Hence it should either be run into distinct tanks and treated especially, without being allowed to mix with the fresh sewage—which involves extra expense—or it must be let flow over a portion of land. Above all things, it must not be allowed to escape into the river, as I have seen done.

The press cakes, coming from the filter press, and containing about 50 per cent. of water, may be further dried in various manners. They may be loosely stacked on racks in a shed, freely open to the wind, but secure from rain. In this manner the moisture may be brought down in time to 20 per cent.; or the cakes may be shot upon drying floors, heated by suitable furnaces below, or by means of steam pipes, or hot-air pipes.

I have seen the press stuff moulded into a kind of brick, a little larger than an ordinary brick. These are then exposed to air and sun, just as are bricks before being placed in the kiln; or they may be dried at temperatures not exceeding 80° F., by means of waste heat. I have seen sewage deposit dried in this manner down to $13\frac{1}{2}$ per cent., without the addition of lime, gypsum, ashes, or rubbish of any kind.

Perhaps the best arrangement for drying, generally speaking, is the drying cylinder, patented in 1872 (No. 314), by Gibbs and Borwick, a section of which is shown in the accompanying figure. The cake stuff is mechanically agitated in an iron cylinder, whilst a current of hot air passes over it, at the same time. The fumes given off, which, though not hurtful, are unpleasant, may be passed through a scrubber or a coke tower, and finally into the chimney of the engine furnace.

After its passage through the drying cylinders the manure retains from 30 to 35 per cent. of water. But if allowed to lie in heaps in a shed, it loses the greater portion of this moisture, down to about 15 per cent., in which condition it is usually sold.

As a matter of course, if the sewage deposit is to be applied close at hand, so that the extra weight carried is little object, or where there is direct water carriage to the place of consumption, the stuff from the cylinders, or even the press cakes, may be dispatched at once.

CHAPTER XVII.

SEWAGE MANURES.

THE question of sewage manures and their value, is one which has been very needlessly complicated by extraneous considerations. We know now that plants require food, just as do animals; they cannot create this food, but can merely appropriate such matters as are necessary for their growth. These matters they take up, partly from the air, partly from the rain, but chiefly from the soil.

Now the supply of plant food in the soil is by no means infinite. If we plant crops and reap them year after year without returning anything to the soil, it will ultimately become exhausted. Plants will grow in it less and less luxuriantly than they did at first, and by-and-by they will cease to grow at all, and the land will be rendered barren. Millions of acres in the countries bordering on the Mediterranean have been brought into this state. Abundance of land in the United States, once exuberantly fertile, can now be scarcely cultivated at a profit. We said, "Without returning anything to the soil." But if we do make a full and due return, if we give back to the land all the residues and waste of the crops which have grown upon it and all the excrements, liquid and solid, of the animals which have directly or indirectly been fed upon such crops, the land will remain substantially at its original point of fertility.

This has been attempted more or less successfully in all old peopled countries, where land is scarce. In China and Japan the attempt has been successful, because the whole of the excretions have been returned to the soil. In most European countries, and especially in England, the restitution has been very incomplete, because the excretions of the human inhabitants have not been brought back to the soil, but, especially since the introduction of water closets have been run into the rivers, thence into the sea, and thus wasted. Hence, the deficiency has had to be made good, either by the aid of mineral matters mined in this country—coprolites—or by means of matters imported from abroad, such as guano, apatites, kainite, bone ashes, etc.

Now it is very plain that if a country, in addition to sending abroad its agricultural produce, exports manures also, such especially as bones, bone charcoal, bone ash, and dried blood, its own soils must before long become exhausted. Hence we find that countries which formerly exported such products to England have now ceased so to do, the home demand having so far raised the price that the transaction is no longer profitable.

Other and more distant regions will gradually, as their population becomes denser, follow in the same track. Ultimately, they will even cease to export agricultural produce, and will compete with us for a supply of mineral plant foods, nitrate of soda, apatite, phosphorite, and the like. Thus, ultimately, every country will be driven back upon the produce of its own soil for support. The bearing of this truth upon the sewage question is most evident. Will it in the coming time

be at all practicable to go on wasting the excreta of
our urban population as we are now in most cases still
doing? Shall we not want it to keep up the fertility
of our own lands against the time when the foreign
supply, both of food and manures, begins to show a
marked decline?

By many of our present systems, viz., Bazalgettism, intermittent downward filtration, the permanganate process, the Scott cement processes, etc.,
we are exhausting the food producing capacity of any
and every part of the earth which supplies us with
food.

Suppose that London is fed on Indian wheat,
Australian mutton, and Argentine beef. Be it so: we
are then busy sterilising India, Australia, etc., and let
this game be carried on long enough, and generally
enough, and the whole world will become comparatively
barren. Surely, therefore, we should be wise in time,
and desist from our profligate waste.

There is a further consideration; it is commonly said
that the money spent in making sewage manures is not
well laid out, and that plant food might be had at a less
cost from other sources. *For the present* this may be
true. But on sanitary grounds we are compelled to
purify the sewage; and the question is therefore whether
we shall do this in pure waste, or whether we shall not
recover, at least, a part of our outlay in the form of
manure? To a practical mind there can be here no
room for doubt.

Sewage manures, it must further be remembered, are
by no means so deficient in the most essential constituents of plant food as some would-be authorities
wish the public to suppose. Average samples of the

Aylesbury sewage manure, taken absolutely at discretion, have been lately analysed by Professor Dewar and Dr. Tidy, and found to contain 3 per cent. of available ammonia and phosphoric acid equal to about 5 per cent. of tricalcic phosphate of lime. If we remember that farmyard manure in its ordinary condition contains only about 0·5 per cent. of available ammonia, we shall see that sewage manures are not necessarily to be despised, even from a mere analytical point of view. If we look at the results obtained in agricultural and horticultural practice, results obtained from so many and so different quarters as to eliminate all possibility of bribery, corruption, or mistake, we shall find them much better than the mere percentage composition of the manure would lead us to expect. Hence, we are pointed to the truth that the condition, as readily assimilable or otherwise, of a manure, may be of as great importance as the mere percentage of its constituents.

There are, however, many different grades of sewage manures. If the sewage of a town consists largely of industrial waste waters, the matter precipitated or absorbed from it may have but a very low agricultural value, acting *worse* in practice than its analytical value would lead us to expect. Such sewages, it must be remembered, are sure to give unsatisfactory results if applied in irrigation, certain of their constituents being more or less poisonous to plants. But when the sewage of a town brings with it the entire excreta of its human and animal inhabitants, *plus* the blood from the slaughter-houses, &c., and all this dissolved or suspended in a volume of water, not greatly exceeding 30 gallons daily per head of the population, and with little or no dye or tan liquors, solutions of metals, etc., such sewage, if properly treated, will yield a manure of approximately the strength above mentioned.

The manner of treatment, of course, greatly affects the result. The lime processes not merely throw down from the sewage a smaller proportion of the dissolved organic impurities than do, *e.g.*, the salts of alumina, but the lime goes on exerting a decomposing action upon the sediment, expelling gradually the ammonia, and thus constantly reducing its manurial value. Other changes take place simultaneously, as may be judged from the peculiar sickly smell which lime deposits give off for a very considerable time.

Further, though lime, in the caustic state, or as carbonate, is very commonly applied as a dressing to land, and often with good results, yet it is never used along with any organic manurial matter, whether farmyard manure, night-soil, guano, refuse fish, etc. On the contrary, the lime is applied by itself, generally in the autumn, after the crops have been gathered, and the nitrogenous manure is used some months later, in the beginning of spring. But if the deposit from a lime sewage process is applied, we have the very mixture which practice has taught the farmer to avoid. These remarks will hold good whether the sewage has been precipitated with lime by itself, with magnesian lime, or with lime in conjunction with phosphates, with copperas or other salts of iron, salts of alumina, etc. In short, we may say that no precipitate from an alkaline effluent is likely to prove very satisfactory as a manure.

It need scarcely be said that sewage deposits thrown down by any poisonous agent, such as gas lime, salts of baryta, zinc or lead, are quite inadmissible as manures.

Gas lime and alkali waste precipitates may, however, be usefully applied to kill weeds, heath, etc., on waste lands which it is desired to bring under cultivation. Even in such cases, it might be simpler to apply the gas lime, etc., at once, without first passing it through sewage.

Very powerful oxidising agents, such as the manganates, the permanganates, and the chromates (if the latter should ever come into use), must, *pro tanto*, lessen the value of the manure.

The deposit from sewage left to itself in a settling tank, without chemical treatment, is worthless. It contains very little of the dissolved organic matters of the sewage, which during the settling process have taken to themselves wings and flown away in the form of noisome gases, often conveying putrefactive and disease engendering organisms. It consists chiefly of the sand, gravel, comminuted stones, etc., which the sewage sweeps along, especially in towns where the streets have been dealt with according to the iniquity of Macadam.

At a certain sea-coast town a curious mistake led to the production of a sewage precipitate—supposed—very similar in character and value to the spontaneous deposit just mentioned. The precipitation tanks were so situate that the effluent water could at low tide be run off into the sea. The manager of the works, instead of adding the precipitating and occluding agents to the sewage, conceived the original but most unhappy thought of allowing the sewage to subside spontaneously, running the liquid off into the sea and adding the precipitating mixture to the deposit. No good end was effected by this strange process; the organic impurities which should have re-appeared in the manure, passed out to sea. To make the matter worse, the town in question lies, in part, on a very steep declivity, so that in time of heavy rain the surface water rushes furiously down to the lower levels charged with little but silt. This material, after settling in the sewage tanks, and being sprinkled over with the precipitating mixture, was sold as manure to a very considerable extent before the fatal blunder was detected. The consequence is that in the districts around the town in question, comprising some of the most highly cultivated land in all England, sewage manure, no matter from what source or of what quality, has been brought into utter disrepute and would scarcely be accepted as a gift. I had the opportunity of observing a large tract of meadow land, one half of which had received a heavy dressing of this supposed sewage manure, whilst the other had received no manure at all. On examining the land carefully I was unable to perceive any difference between the manured and the unmanured portion, and this in the beginning of June.

I cannot help noticing that a part of the unmerited contempt and neglect with which sewage manures are greeted is due to the efforts of the manufacturers of super-phosphates, dissolved guano, and other chemical fertilisers. These gentlemen very naturally wish to keep the market to themselves, and of course object to an increased supply, and an improved quality of sewage manures, as an unwelcome competition. Thus we find one of the most eminent manure manufacturers as above noticed, arguing in favour of Bazalgettism, and pronouncing it good policy to run our sewage into the sea to feed the fishes. Of all men, super-phosphate makers are the very last who should be consulted on sewage treatment, since their own interests lead them to favour the waste and the destruction of excreta.

CHAPTER XVIII.

SEWAGE LEGISLATION.

UNDER this head we may briefly consider the laws relating to the pollution of rivers, water-courses, wells, and springs by household sewage, and of the waste waters of manufacturing establishments, or of any other source of pollution. Previous to 1876, there was no special statute bearing upon the subject. A riparian proprietor, aggrieved by the fouling of a stream passing by or through his estate, might apply to the Chancery Division of the High Court of Justice for an injunction restraining the offenders from continuing the pollution. Such applications were frequently successful, and they appear to have been among the first causes which led to attempts for the purification of sewage and waste waters.

But the right to turn refuse, solid and liquid, into a river—in other words, to pollute such river—seems in some cases to have become included by prescription among the easements of an estate situate on the banks. At least, at a meeting on the pollution of rivers, one of the most eminent calico printers in Great Britain, stated that some years previously he had begun experimenting on the purification of the waste waters turned out from his works. To his great surprise he soon received a formal letter from the ground landlord, warning him that by so doing he was imperilling one of the prescriptive rights of the estate, and consequently violating one of the covenants of his lease! Thus in certain cases river pollution was not only facultative, but was even a duty.

Then came the Act of 1876, a mild measure, which has been a failure because it has not been duly put in force. It did not contain any fixed standards for the quality of the waters which might and might not be legally allowed to flow into rivers. It ignored the well-known " Recommendations" which have been discussed in a previous chapter. Nor was there any attempt to force some particular method of sewage treatment upon the nation. Whatever defects the Act might possess— and to two of them we are about to turn our attention —its framers were thoroughly aware of the delicacy and difficulty of the task. Whilst acting in the spirit of the grand maxim, "*sanitas sanitatum et omnia sanitas*," they were careful not needlessly to harass municipalities and manufacturers. They wished to effect the desired reforms gradually, not laying down at once hard and fast lines, which could only be enforced by an offensive espionage, by continual fines, and by strengthening the hand of an aggressive bureaucracy.

But the Act did not even attempt to codify the law on the pollution of rivers. It is what, I believe, is called in the language of the law a " cumulative " measure, that is, a statute which provides new procedures and new penalties for certain omissions and commissions, but leaves all old precedents, customs and statutes, bearing upon the particular subject, still perfectly valid. A riparian proprietor who knows or fancies that some manufacturer or municipal authority is polluting a river has under that Act still a choice of procedures. He may take action under the statute, or he may, I suppose, still apply to the Chancery Division for an injunction, just as if the Act had never been passed.

Now, this power of option, with all the uncertainty hanging over the subject, may be very convenient for a litigious person, who, from whatever motive, wishes to harass a neighbour. But it is, I submit, grossly unfair to municipalities and to manufacturers. Both are surely, in common fairness, entitled to have open to them a clear, definite, comprehensive statement of what is legal, and what is illegal, as regards the disposal of sewage and waste waters, so that they may know when they are obeying the law, and may feel assured that when so doing they are not open to *any proceedings whatever*.

On the other hand a riparian owner, a municipality, or manufacturer, situate down stream, and feeling aggrieved by the actions or omissions of a neighbour up stream, might surely be satisfied if one sufficient and efficient remedy is placed at his disposal.

Another shortcoming of the Act of 1876 and of all existing law on the subject has been brought to light very lately. The law provides remedies in case of the pollution of any river, brook, lake, or other piece of water *open to the air and light*. But concerning underground waters it is silent. An explanatory instance is here necessary. A. B. had within the limits of his property a well which supplied his household with potable water. On the property of his neighbour, C. D., there was also a well, disused as a water supply. C. D. took upon himself to turn the sewage of his house into the abandoned well. Before long the various kinds of polluting matter found their way through the intervening subsoil and contaminated A. B.'s well. The latter thereupon took proceedings for the abatement of so grave a nuisance; but the judges, finding neither statute nor precedent bearing upon the question, were compelled to decide in favour of the defendant, or offender.

In 1885, we had an attempt, though an unsuccessful one, to improve the law of river pollution. Earl Percy, Colonel Walrond, and Mr. Hastings brought in a Bill of a decidedly more drastic character than the Act of 1876. As a prominent feature it embodied the very objectionable "Recommendations" of the late Royal Rivers' Pollution Commissioners, which we have already discussed. It adopted also their cardinal errors, of enforcing one standard for all the rivers in the kingdom, of paying no regard to the river itself, and of looking merely at the *quality*, neglecting the *quantity* of any polluting influx.

The Bill also made no attempt to remedy those serious defects in the Act of 1876 which have just been pointed out. There is assuredly no provision against the pollution of underground waters. Codification further, is distinctly repudiated. It is laid down that: "The powers given by this Act shall not be deemed to prejudice or affect any other rights or powers now existing or vesting in any person or persons by Act of Parliament, law or custom; and such right or power may be exercised in the same manner as if the Act had not passed, and nothing in this Act shall legalise any act or default which would, but for the Act, be deemed a nuisance." This is one of those "saving clauses" to which an irreverent friend of the writer's applies an exactly opposite epithet.

Not having enjoyed a legal training I cannot presume to form an opinion as to whether, in virtue of this clause, any person who had enjoyed by prescription the right of polluting some river, would have been justified in continuing the pollution.

Several most mischievous clauses disfigured the Bill. There may be, and probably will be found, cases where the river is worse than the sewage or waste waters discharged into it. For such cases the Bill provided that:—" It shall be no defence to any offence against this Act, to prove that, after any offence against the Act has been committed, the water of the stream is less polluted than is defined by the standard of purity hereinafter mentioned." Had this Bill unhappily become the law of the land, a manufacturer might have incurred heavy penalties—£50 per day—for turning into a stream water purer than the stream itself, if only such water transgressed the standards in any one point. Yet such a manufacturer, instead of polluting the river would be improving it. Further still: if a stream is in any one point less pure than required by the standards it would have been an offence to take out a quantity of such water for any purpose, and let it flow back in its original condition.

CHAPTER XIX.

SEWAGE PATENTS.

Abstracts of Specifications for the Chemical Treatment of Sewage.

1. *W. Higgs.* 1846. *No.* 11,181.
   Precipitates with slaked lime, and brings the sewage gases evolved in contact with chlorine or hydrochloric acid gas.
2. *J. H. Browne.* 1850. *No.* 13,280.
   Precipitates with basic salts, such as subsulphate of peroxide of iron, or any neutral metallic salt, or metallic salts mixed with cheap oleaginous matter. Or chloride of calcium.
3. *T. Wicksteed.* 1851. *No.* 13,526.
   Uses milk of lime.
4. *R. Dover.* 1851. *No.* 13,775.
   Treats with hydrochloric acid or other mineral acid, iron filings, oxide of iron, chloride of sodium or protosulphate of iron. He then filters effluent through charcoal, clay, gypsum or peat.
5. *H. Stothert.* 1852. *No.* 14,073.
   Precipitates with caustic lime, sulphate of alumina, sulphate of zinc, compound animal and vegetable charcoal (obtained by distilling the precipitated matters of sewage waters, or by distilling night-soil, creosote or oil of peat), peat mould, tanner's spent bark, burnt

clay, old mortar or mixture of such matters or other matters.

6. *A. W. Gilbee. New Act*, 1852. *No.* 250.

Applies to drains and sewers a deodorising powder, obtained by burning lignites, or any ligneous substance, etc.

7. *W. Bardwell.* 1853. *No.* 29.

Filters within a close chamber, and receives the gases in suspended trays of sawdust moistened with sulphuric acid.

8. *J. F. Pinel.* 1853. *No.* 581.

Adds to sewage sulphate of zinc, potash, alum, common salt and sand (!)

9. *J. T. Herapath.* 1853. *No.* 643.

Precipitates the phosphoric acid and ammonia of sewage by the addition of magnesian compounds at or about the same time as the addition of some chemical agent which will not decompose ammonia or its salts, but will combine with or absorb hydrosulphuric acid, such as metallic sulphates or metallic chlorides, or animal or vegetable carbon.

10. *T. T. Dimsdale.* 1853. *No.* 1,252.

Adds to the sewage a peculiar kind of peat-earth containing a salt or salts of iron or oxide of iron.

11. *A. Macpherson.* 1853. *No.* 1,511.

Mixes with the sewage peat charcoal, mixed with common salt.

12. *J. A. Manning.* 1853. *No.* 2,780.

Treats sewage with animal charcoal, alum, carbonate of soda and gypsum.

13. *A. Macpherson.* 1853. *No.* 2,876.

Uses peats dried or charred, charred saw dust, sulphuric acid, common salt, hydrate of lime, quicklime, brick-dust, or dry clay as a filter bed.

14. *A. R. Smith and A. Macdougall.* 1854. *No.* 142.
Treat sewage with magnesia and lime, combined with sulphurous acid or carbolic acid.

15. *T. Wicksteed.* 1854. *No.* 193.
Precipitates with lime and finely divided charcoal, preferably prussiate carbon.

16. *T. J. Herapath.* 1854. *No.*
Uses coke of Boghead coal for drying, deodorising, or absorbing sewage.

17. *J. A. Manning.* 1854. *No.* 709.
Treats sewage with "soft sludge" from alum works.

18. *J. Littleton.* 1854. *No.* 2,532.
Separates gases from sewage by the action of a vacuum, and absorbs them in charcoal or other matter.

19. *J. A. Manning.* 1855. *No.* 1,786.
Treats sewage with alum-slate, alum-shale, alum-schist, alum-stone, alum-ore, and other aluminous minerals. He uses also powdered lime and animal charcoal.

20. *T. Wicksteed.* 1856. *No.* 1,815.
A mere filtration arrangement.

21. *J. T. Victor.* 1856. *No.* 2,273.
Uses as an "anti-putrefactive," sulphate of zinc and copper and water.

22. *G. Vertue.* 1856. *No.* 2,842.
Deodorises sewage with quicklime, taken hot from the kiln, slaked close to the sewers and passed into them.

23. *F. H. Maberley.* 1857. *No.* 49.
Mechanical arrangements for separating the solid from the liquid portions of sewage.

24. *Sir James Murray.* 1857. *No.* 114.
Appears to treat sewage with carbonic acid, generated and applied in various manners.

25. *H. Medlock.* 1857. *No.* 186.

Purifies water by contact with bundles of iron wire, and afterwards filtering.

26. *J. W. Rogers.* 1857. *No.* 992.

Filters through peat-charcoal.

27. *J. Lloyd.* 1857. *No.* 1,793.

Treats sewage with ashes and cinders, mixed with lime, and occasionally with chloride of lime.

28. *F. Lipscombe.* 1857. *No.* 2,168.

This patent is mentioned here merely as containing the germ of the Bazalgette process. The sewage is to be run through iron culverts into the sea.

29. *W. T. Manning.* 1857. *No.* 2,949.

A filtration process.

30. *J. A. Manning.* 1858. *No.* 61.

Treats sewage with refuse from chlorine stills, *i.e.*, manganese chloride.

31. *G. Niellon.* 1858. *No.* 171.

Sewage is led into reservoirs and is assumed to be precipitated, though no agents are mentioned. The effluent is filtered through vegetable matter.

32. *J. A. Manning.* 1858. *No.* 179.

Collects sewage in tanks, and precipitates the solids with alum-sludge, or its chemical equivalents (see Specification No. 1,786, 1855), and combined, if necessary, with refuse charcoal, lime, or *chloroform* (*sic*) !

We can only hope, in common charity, that "chloroform" is a clerical or typographical error.

33. *G. L. Blyth.* 1858. *No.* 287.

Employs a soluble phosphate to neutralise and fix any ammonia or nitrogenous matter in sewage, or other fluids, such as gas, or dye, or other waste liquors, and also to neutralise the excess of acid by lime, or other alkali, or alkaline earth, or magnesia, or magnesian limestone, or aluminium, so as to precipitate the whole of the phosphate from the liquid. Instructions are given for preparing soluble phosphates, and in place of phosphate of lime, phosphate of magnesia, alumina, iron, or copper may be used, as these, except phosphate of magnesia, which does not require the addition, can be brought to the soluble form, and by double decomposition with a soluble salt of magnesia converted into superphosphate of magnesia.

In most dye-liquors nitrogenous matter is not abundant.

34. *J. Hadfield.* 1868. *No.* 400.

Filters sewage through a bed, the first layer of which is spent tanner's bark, saturated with solution of copperas; the second is of chalk; the third again of vegetable matter, saturated with copperas. The liquid is then again filtered through chalk, or a mixture of chalk and sulphate of baryta.

35. *T. Spencer.* 1858. *No.* 1,415.

Filters foul waters through a bed of magnetic iron carbide.

36. *B. Young and P. Brown.* 1,858. *No.* 1,460.

Let sewage settle in tanks, pass gases through furnaces, and after some days allow supernatant water to flow into river or sea, and apply the solids usefully.

37. *J. Chisholme.* 1858. *No.* 1,499.

Treats sewage by electricity, for doing which he describes eight processes.

38. *A. R. Broonan.* 1858. *No.* 1,616.

Precipitates the solids of sewage with lime water or soluble salts of lime. Mixes solids with gypsum recently baked to plaster.

39. *J. B. A. Duglere.* 1858. *No.* 2,073.

Separates solids from liquids for disinfecting purposes, by employing magnesia, salts, metallic precipitants and phosphoric and ammoniac acids. What "ammoniac acid" may be is doubtful.

40. *H. Moule.* 1859. *No.* 539.
Treats sewage by evaporation.

41. *Bridge Standon.* 1859. *No.* 1,567.
Treats sewage and excrement with mineral oil, paraffin, gallipoli or rape oil; common salt may also be applied. Sulphuric acid is used to neutralise ammonia.

42. *N. Heckford.* 1859. *No.* 2,107.
Proposes to purify polluted rivers by forcing sea water up them, and applying it also to cesspools.

The inventor was probably not aware that the contact of polluted streams with sea water is exceedingly dangerous to health.

43. *John Dales.* 1859. *No.* 2,157.
Treats sewage with magnetic-chloride of iron, "prepared by dissolving in muriatic acid magnetic oxide of iron, or any of those ferruginous products from various metallurgical processes and manufactures which are magnetic." He uses also chloride of manganese.

This approaches very closely to a solution of iron recently sold in France and Italy as a disinfectant, under the name of "Smorbo."

44. *R. Smith.* 1859. *No.* 2,359.
Treats sewage with a mixture of powdered sulphur and turpentine, "with any of the ordinary agents employed to animal or vegetable matter in a state of decomposition."

45. *A. McDougall.* 1859. *No.* 2,958.
Treats sewage with a heavy oil of tar, dissolved in water, by the use of alkali, or alkaline earths; and secondly, treats the oil of tar with an acid previously, to increase its solubility. If the sewage is already putrid

a small quantity of a metallic salt may be added to the oil, after treatment with the acid.

46. *M. Mangles.* 1860. *No.* 655.

Forces chlorine, or sulphurous, nitrous, hydrochloric gases through sewage, &c.

47. *J. A. Manning.* 1860. *No.* 1,343.

Alum-sludge again; but the sewage is allowed to settle first for some time, and only the clear water is to be treated.

48. *F. Durand.* 1860. *No.* 1,612.

Constructs a canal to the sea, with gates to be open or closed, according to the state of the tide.

49. *J. Harrop.* 1863. *No.* 132.

Treats fæcal and urinous matter with chloride of sodium alone, dissolved in water, or mixed with suitable resinous, bituminous, earthy, vegetable, metallic, or alkaline matters.

50. *W. Clark.* 1863. *No.* 761.

Filters through a layer of "crotels" sulphate of magnesia in powder, on other layers of "crotels," saturated with free phosphoric acid, and phosphate of lime. As other filtering materials are mentioned "fæces and paunches of animals" (!).

51. *R. C. Clapham.* 1860. *No.* 976.

Treats sewage with chlorides of manganese and iron.

52. *W. Clark.* 1863. *No.* 1,362.

Passes urine, &c., upon beds of quick-lime.

53. *H. Martin.* 1863. *No.* 1,435.

Mixes sewage or night soil with charred tan, or other charred vegetable material.

54. *T. H. Baker and G. Friend.* 1863. *No.* 2,208.

Filter through sand, shingle, waste tan, gravel, charred tan, &c.

55. *F. Maynes.* 1863. *No.* 3,264.

Freezes sewage, &c., to effect separation of the solids.

Even if this process were practicable it would not fully remove either dissolved or suspended impurities.

56. *W. E. Weston.* 1864. *No.* 1,688.

Treats sewage with neutral sulphate of peroxide of iron (ferric sulphate).

57. *W. Bardwell.* 1864. *No.* 3,115.

Precipitates the solids of sewage with sulphate of lime.

58. *H. Bird.* 1864. *No.* 3,160.

Mixes sewage waters with sulphated clay, *i.e.*, crude aluminium sulphate.

59. *R. Smith.* 1865. *No.* 451.

Introduces into the sewers peat, saw-dust, or other ligneous matter saturated with sulphurous acid, or superphosphate of lime.

60. *C. G. Lenk.* 1865. *No.* 1,242.

Adds to foul water, a mixture of permanganate of potash, carbonate of soda, alum and neutralised aluminite.

61. *A. Bird.* 1865. *No.* 2,415.

Precipitates with sulphate of alumina.

62. *J. Linton.* 1865. *No.* 2,626.

Adds to sewage solids ground clay, or other suitable refuse, with a small quantity of gypsum (!) as a deodoriser.

63. *C. G. Lenk.* 1865. *No.* 2,674.

Adds to foul waters alum, "alumnite" neutralised or not, two parts carbonate of soda and solution of iron, or of permanganate of potash.

This border very closely upon No. 1,242 of the same year.

64. *H. Y. D. Scott.* 1865. *No.* 2,808.

Deodorises sewage with lime and metallic salts, is chiefly perchloride or persulphate of iron, the former to neutralise the phosphoric and carbonic acid contained in sewage, and the latter to seize upon the sulphuretted hydrogen. The patentee admits that these substances have been already used, but not in the right way ! He causes the lime to act upon the sewage in a separate reservoir and then runs off the effluent into another tank, where it is treated with the solution of iron.

It is only necessary to remark that the neutralisation of carbonic acid is a point of no importance, and that in recent sewage sulphuretted hydrogen is not necessarily present. The patentee adds also sometimes sulphate of lime, common salt, and carbolic acid.

65. *F. Sutton.* 1866. *No.* 101.

Treats sewage with sulphate of alumina, alone or in conjunction with clay, sulphate of magnesia, or peroxide of iron.

66. *G. E. Moone.* 1866. *No.* 1,163.

Treats with lime and distils off the ammonia which is received in hydrochloric acid, forming sal-ammoniac.

67. *A. Kühne.* 1866. *No.* 2,107.

Treats foul water with chlorine or alkaline permanganates, with or without sulphate of iron or other metallic salts ; if excess of permanganate has been used, the inventor neutralises it with hypo-sulphite of soda.

68. *R. Irvine.* 1866. *No.* 2,218.

Treats sewage with mineral charcoal obtained from the residuum left by the distillation of paraffin oil from shale, and containing silicate of alumina, with lime, magnesia, oxides of iron and carbon in a finely divided state.

This specification should be carefully compared with that of Mr. Coleman, of Glasgow, No. 1,954, of 1875.

69. *A. H. Bonseville.* 1866. *No.* 2,926.

Treats foul waters with lignite *coke*, charcoal, sulphate of iron, potter's clay, and slaked lime.

70. *Ernst Süvern.* 1867. *No.* 119.

Treats sewage, &c., with a mixture of burnt lime, with 5 per cent. of coal tar, and about ten times as much water as lime, with the addition, when necessary, of 10 to 25 per cent. chloride of magnesium. The reader is requested to compare this specification with the subsequent ones, 1870, No. 3,167; and 1876, No. 1,355.

71. *W. Parry and J. Frearson.* 1867. *No.* 417.

Mix with the sewage clay, clay iron ore or manganesic earths, and allow it to settle.

72. *A. H. Hart and W. Parry.* 1867. *No.* 788.

Use the same materials.

73. *E. Guenin.* 1867. *No.* 1,229.

Uses waste manganese chloride from the chlorine stills, neutralising acid if needful with dolomite, or other calcareous magnesian substance, or lime, or zinc or iron scrap, or oxidised ores of the same. He also adds 2 or 3 per cent. of raw salts of alumina from the washing of aluminous schists, or a certain quantity of the schists themselves in a natural state.

74. *F. Tolhausen.* 1867. *No.* 2,549.

Treats urine with plaster, peat, and ashes.

75. *T. H. Baker and T. Woodroffe.* 1867. *No.* 2,894.

No purifying agent is mentioned beyond baked earth.

76. *A. M. Clark.* 1867. *No.* 3,566.

Treats sewage with neutral phosphate of magnesia, in order to precipitate ammoniaco-magnesian phosphate.

77. *W. C. Sillar, R. G. Sillar, and G. W. Wigner.* 1868. *No.* 1,954.

Treat sewage with 4 lbs. per 1,000 gals. of the following mixture:—

	Parts.
Alum	600
Blood	1
Clay	1,900
Magnesia	5
Manganate of potash	10
Burnt clay	25
Chloride of sodium	10
Animal charcoal	15
Vegetable charcoal	20
Magnesian limestone	2
	2,588

This is the original A B C process. It may be sufficient here to remark that the chloride of sodium is certainly injurious, whilst the magnesian limestone, the burnt clay, and the magnesia are, under most circumstances, inert.

78. *W. H. Hughan.* 1868. *No.* 2,883.

Mixes the sewage with cement to the consistency of mortar. Acids or salts may be added, *e.g.* 1 part of copperas to 6 parts of cement. Describes a special cement made of 4 parts alum and 1 part of clay or sulphate or phosphate of lime in solution; a little caustic lime or phosphates may be added, as also charcoal or salt, also bone ash.

It is difficult to conceive the quantity of cement which would be needed to solidify the sewage of London.

79. *E. H. Prentice.* 1868. *No.* 2,919.

Adds to sewage phosphoric acid, or any soluble phosphate, in the proportion of 12 or 15 lbs. to 1,000 gals.

sewage, and then precipitates with 20 to 30 lbs. lime to 4,000 gals. sewage.

This is a favourable specimen of the phosphate process.

80. *G. Chapman.* 1868. *No.* 3,203.

Lets the sewage decompose in large tanks (!) at a temperature of 70 to 80 degs. Fahr.; then adds caustic lime in a precipitating tank, and passes through the clear liquor steam to extract the ammonia.

The first part of this process would be a fearful nuisance. A part of the ammonia would be lost on addition of the lime, if not before.

81. *C. Jones.* 1868. *No.* 3,457.

Precipitates sewage with slaked lime and petroleum, or, instead of petroleum, the acid tar obtained in the manufacture of liquid hydro-carbons. To improve the manure, chloride of zinc, or sulphuric acid, or burnt clay, or magnesia may be added.

What is to precipitate the petroleum? Chloride of zinc is a poison to vegetation.

82. *T. Smith and J. van Norden Bazalgette.* 1868. *No.* 3,562.

Treat sewage with a mixture of marl, clay, mould, schists, refuse products, ashes, treated by mineral acids.

The term refuse products includes substances some of which would be useless, and others positively injurious.

83. *A. M. Clark.* 1868. *No.* 3,714.

Treats sewage with a double phosphate of magnesia and iron.

84. *M. J. Anderson.* 1869. *No.* 3,550.

Precipitates with sulphate of alumina (1 lb. to 100 gals.) followed up by 5 lbs. slaked lime.

This patent became the property of the "Rivers Purification Association," and has been worked for some years at Coventry and Nuneaton, followed up, however, by irrigation. It will be observed that the lime is much more than sufficient to neutralise any acids likely to be present in the sewage, and will therefore act as a substantive precipitant.

85. *W. H. Hughan.* 1870. *No.* 67.

Uses to sewage natural phosphates, treated with dilute acids, diluted with urine and mixed with night-soil, along with the cement indicated in his former patent, 1,868, No. 2883.

86. *D. Forbes and A. J. Price.* 1870. *No.* 607.

Add to the sewage phosphate of alumina, previously dissolved in sulphuric acid, and follow up with lime. The proportion of phosphate preferred is 2 lbs. to 1,000 lbs. of sewage.

This process, like the two following, came into the hands of the Phosphate Sewage Company, and was worked at Hertford.

87. *D. Forbes and A. P. Price.* 1870. *No.* 1,137.

This specification differs very little from that of 1870, No. 607. The inventors now propose to heat the phosphate of alumina in hydrochloric acid or in a mixture of that and sulphuric acid, and they also add deodorising agents, such as animal or vegetable charcoal.

88. *A. P. Price.* 1870. *No.* 1,314.

This invention, again, borders very closely upon Nos. 607 and 1,137 of the same year. The inventor uses "natural phosphates of iron, lime, and alumina."

89. *G. W. Wigner.* 1870. *No.* 1,354.

This is an improvement upon No. 1,954 of 1868. The inventor uses :—

	Parts.
Alum	600
Blood	1
Clay	1,900
Magnesia	5

	Parts
Manganate of potash	10
Burnt clay	25
Chloride of sodium	10
Animal charcoal	15
Vegetable charcoal	20
Magnesian limestone	2
Sulphate of alumina	169
Sulphate of iron	3
Sulphate of lime	66
Alumina	94
	2,920

There is added the proviso that, instead of the last four substances, 488 parts of crude alum may be used, making the entire quantity of alum to be used 1,088 parts.

It is certainly strange to find alum and sulphate of alumina combined in the same formula. Alum is not only much more costly and less readily soluble, but has the serious disadvantage of introducing into the effluent water, in pure waste, sulphate of potash or sulphate of ammonia. The useless ingredients of the patent No. 1,954 of 1868 (magnesia, burnt clay, chloride of sodium, and magnesian limestone), are retained, and two new ones, sulphate of lime and alumina (unless the hydrate be meant), are introduced. This patent came into the possession of the Native Guano Company, but it has long been abandoned in favour of simpler and more rational processes.

90. *B. G. Sloper.* 1870. *No.* 1,706.

Mixes fresh sewage with stale sewage to promote fermentation. Then draws off the supernatant water (!), and mixes with the sediment a salt of magnesia and phosphate of soda or lime. To deodorise the sewage he uses sulphate of alumina, sulphate of iron, sulphate of lime, and chloride of lime.

To the fermentation process there is a strong objection, mentioned under No. 3,203, A.D. 1868. Sulphates of alumina and iron precipitate, but they have little deodorising power. Sulphate of lime is altogether out of place in sewage treatment, and chloride of lime, if used in sufficient quantity to deodorise sewage, will destroy fish in any river into which the effluent may make its way.

91. *J. J. Hays.* 1870. *No.* 2,297.

Treats sewage with "ground peat, peat charcoal, or other suitable material." Lets settle, and filters effluent.

92. *G. Bischof.* 1870. *No.* 2,516.

Passes sewage upwards or downwards through a layer of spongy iron, about a foot thick, at the bottom of a tank or filtering bed. The solids are to be previously filtered out.

This process, which involves double filtration is, of course, inapplicable when the sewage contains any substance capable of acting on iron.

93. *F. Fenton and S. Hollins.* 1870. *No.* 2,534.

Treat sewage with a combination of gypsum, sulphate of lime, bisulphate of iron, soot, chalk, salt, cinder breeze or ashes, or a combination of the bind, clay shale, alum shale, or barren shale of the coal measures, or alum clay or alum shale, either raw or roasted with any or all of the above.

For acid waters they use any of the above ingredients, or salt and lime, or salt, lime, gypsum, and chalk, or plaster of Paris, or calcined gypsum, hydrated together or separately, and afterwards mixed. The product may be used "as a substitute for coprolites" (!).

The error of using soot, sulphate of lime, chalk, and salt in sewage treatment has been already explained. Most clay shales, raw or roasted, are quite inert.

94. *C. F. Kirkman.* 1870. *No.* 2,653.

Seeks to disinfect sewage by treatment with carbonic acid, and by passing through a receptacle in which are a number of zinc and copper plates, by which "a continuous current of electricity is made to pass through the sewage."

95. *J. J. Hays.* 1870. *No.* 2,838.

Treats sewage with peat charcoal, and filters through a bed of peat charcoal, and afterwards through carbonate or hydrate of lime and sand. (See 1870, No. 2,297.)

It is necessary to remark that the actions of carbonate of lime and of hydrate of lime upon sewage are not alike.

96. *A. Bryant and S. H. Culley.* 1870. *No.* 3,107.

Let sewage deposit in settling tanks, after the addition of deodorising materials, such as carbolic acid, and then filter through sawdust, dried or charred.

97. *F. Hillé.* 1870. *No.* 3,167.

Uses as a disinfectant for sewage, chloride of zinc, calcium and magnesium, lime and gas-tar. (See specification of E. Süvern, 1867, No. 119, in which the use of chloride of magnesium, in combination with lime and coal tar is already mentioned.)

98. *H. Y. Darracott Scott.* 1870. *No.* 3,169.

Treats sewage with lime, preferably gas-lime (!), together with certain metallic salts, suitable for precipitating sulphuretted hydrogen. These agents are to be introduced at different parts in the course of the sewers.

This system is objectionable since the deposits produced may accumulate and putrefy in the sewer. The metallic salts best fitted for removing sulphuretted hydrogen are poisonous, and too expensive.

99. *C. Rawson, P. Ovenden, James Wylde, W. McCree, and H. Hill.* 1870. *No.* 3,399.

Improvements on No. 1,954, A.D. 1868, and on No. 1,354, A.D. 1870. The inventors substitute for blood, "albuminous, albumenoid, or gelatinous substances." They also propose an alkaline mixture to be added if necessary to the sewage, either before or after the ingredients mentioned in the two former patents.

100. *S. Proctor and J. M. Sutton.* 1871. *No.* 297.

Remove large solids from the sewage by mechanical arrangements; add then disinfectants such as carbolic acid, and filter.

101. *G. B. Sloper and F. J. J. Washer.* 1871. *No.* 329.

Treat sewage first with an alkali, to decompose nitrogenous matters and convert them into ammonia; then add salts of magnesia and soluble phosphates to precipitate ammoniacal magnesium phosphate, and complete the process by adding "small quantities of lime with sulphate of alumina, or chloride of lime, or protosulphate of iron."

Treatment with an alkali fails to decompose some possible nitrogenous compounds altogether, and acts upon others only at very high temperatures.

102. *C. Baly.* 1871. *No.* 351.

Treats sewage with charcoal from the manufacture of acetic acid and slaked lime; one part charcoal, two lime to five sewage.

103. *A. P. Vassard.* 1871. *No.* 1,211.

Treats sewage first with a solution of superphosphate and phosphate of soda, then with sulphate of magnesia, sulphate of alumina and sulphate of ammonia (!). A little lime may also be added.

104. *F. Fenton.* 1871. *No.* 1,897.

In addition to the processes given in No. 2,534, A.D. 1870, the inventor forces atmospheric air through the sewage.

105. *E. Taylor.* 1871. *No.* 1,969.

Treats the solids separated from the liquid sewage with

a mixture of chloride of lime, sugar (!) and alum. The same ingredients are apparently to be used also for urine and blood.

106. *J. T. Lupton.* 1871. *No.* 2,140.

Treats sewage with 20 to 40 per cent of carbon, it may be ashes, with a small amount of phosphate of lime.

107. *H. Y. Darracott Scott.* 1871. *No.* 2,243.

Precipitates sewage with quicklime, dries precipitate and calcines it. Uses the calcined precipitate as manure, or as mortar, or as cement. If used as manure, "superphosphate may be manufactured therefrom."

A substance containing so much lime and so little phosphoric acid cannot prove a very profitable material.

108. *J. Banks and W. Walker.* 1871. *No.* 2,495.

Filter after settling, and mix sediment with sawdust, straw, etc.

109. *J. Hurrow.* 1871. *No.* 2,659.

Treats with an iron, salt and an alkali.

110. *J. B. Pow.* 1871. *No.* 2,760.

Treats in first tank with copperas ; then filters through 1, gypsum, magnesian limestone and charcoal; 2, aluminium shale from the lias containing sulphates of aluminium and potassium, with a portion of sulphate of iron combined with vegetable charcoal ; 3, spongy iron, and 4, cocoa-nut fibre or peat charcoal.

111. *F. L. Hahn Danchell.* 1871. *No.* 2,903.

Treats sewage with a mixture of clay or lime, or both with peat, the mass being charred.

112. *A. P. Vassard.* 1871. *No.* 2,926.

Treats sewage first with lime ; then adds to the liquid oxide of barium and biphosphate of lime ; then with chloride of lime and aluminate of soda. Instead of oxide of barium, the sulphide or other soluble salt may be used ; instead of biphosphate of lime ordinary superphosphate or other phosphate. Also other salts of magnesia (!) may be used instead of the chloride, and various alkaline re-agents instead of aluminate of soda.

The objectionable character of barium compounds has been already noticed ; the sulphide (sulphuret) as giving off sulphuretted hydrogen is the worst.

113. *J. Cole and W. Abbott.* 1871. *No.* 2,975.

Precipitate the sewage in a tank, but do not state what is the agent employed.

114. *H. Smith.* 1871. *No.* 2,997.

Reduces the temperature of the sewage and of the air above the tanks, the latter by allowing liquid carbonic acid to evaporate into it.

Even if this process could be carried out on the large scale it would be useless, as water, if freezing, is not freed from either its suspended or dissolved impurities.

115. *J. F. Fahlman.* 1871. *No.* 3,233.

A new mechanical arrangement, where no particular method of disinfecting the sewage is claimed.

116. *J. A. Wanklyn.* 1871. *No.* 3,436.

Obtains ammonia from sewage, driving it off either by heat, or by " a current of air of a suitable temperature," and condenses the ammonia by means of acid in a coke tower.

117. *H. Y. D. Scott.* 1871. *No.* 3,515.

Brings the effluent from, it would seem, any precipitation process in contact with charcoal. When the charcoal is saturated with ammonia it is dried at a gentle heat so as not to expel the ammonia, then heated to redness and the ammonia collected by any suitable means.

The power of wet charcoal to absorb and retain ammonia from liquids is not strikingly great.

118. *F. G. Prange and W. Whitthread.* 1872. *No.* 379.

Treat sewage with a solution of dicalcic phosphate in an aqueous solution of mono-calcic phosphate, with an alkaline earth or alkali, such as lime. If there is an excess of free ammonia, magnesium salts may also be added.

119. *A. M. Clark.* 1872. *No.* 388.

Fixes the ammonia in sewage by means of bi-magnesian phosphate or calcic magnesian phosphate.

120. *J. Robey.* 1872. *No.* 435.

Mixes peat and clay, burns the mixture, and uses it in treating sewage.

121. *Silvester Fulda.* 1872. *No.* 448.

Treats sewage with unslaked lime, sulphate of soda, and, if required, of nitrate of soda, borax, and silica.

The uses of sulphate of soda in sewage treatment are almost as hard to imagine as the circumstances under which nitrate of soda and borax can be requisite. Except the original sewage is strongly acid, the effluent from this treatment must be alkaline.

122. *F. Hillé.* 1872. *No.* 484.

Mixes sewage with chloride of magnesium, then passes it into another tank, where it is treated with milk of lime, forces carbonic acid gas into it, or adds instead a small quantity of perchloride of iron, and filters over charcoal. In hot weather, a mixture of lime and tar may be added in a subsequent tank, or in the mixing-tank. It is hard to see the precise novelty in this process. See the inventor's previous patent, No. 3,167 of 1870.

123. *S. W. Rich.* 1872. *No.* 547.

Lixiviates aluminous schists, adds to the liquor chloride of sodium and evaporates down. In this manner is obtained a crude chloride of aluminium, fit for treating sewage.

124. *W. E. Gedge.* 1872. *No.* 626.

Treats sewage and other ammoniacal liquids in a succession of boilers and passes the ammonia into dilute sulphuric acid.

125. *R. Blackburn.* 1872. *No.* 671.

Screens and strains sewage, and treats the liquid portions chemically (*how*, it is not stated), or uses them for irrigation, or allows them to run into a water-course.

As in several patents, the fact that the "liquid portions" of sewage are at once the most valuable, and, if run direct into a watercourse, the most dangerous, is here overlooked.

126. *H. Y. D. Scott.* 1872. *No.* 849.

Precipitates the suspended impurities with lime, preferably dolomitic; treats the effluent from first precipitating tank with phosphoric acid or a phosphate to precipitate lime; treats then the effluent from the second tank—may be treated with phosphate of magnesia to extract the ammonia. Various things are also mentioned which "may" be done.

127. *Dugald Campbell.* 1872. *No.* 944.

Treats sewage with acid, phosphate of lime, and then adds milk of lime.

128. *S. W. Rich.* 1872. *No.* 1,243.

Converts the alumina and peroxide of iron in burnt shales into sulphates by treatment with sulphurous acid. The process is carried on continuously in a kiln, in which the shale is burnt while sulphurous acid gas (from burning pyrites) are introduced below. The product formed is lixiviated, and the liquor used for treating sewage.

129. *T. Christy.* 1872. *No.* 1,257.

Disinfects foul waters with heavy oils, and then treats them with silica to form a mixture which the inventor terms "silicoid."

130. *F. L. H. Danchell.* 1872. *No.* 1,394.

Treats sewage with animal or vegetable refuse, mixed with loam, clay, phosphate of alumina, lime, carbonate of lime, or phosphate of lime, and charred.

131. *James Robey.* 1872. *No.* 1,421.
   Chars arable soil in retorts, and uses the product for treating sewage.

132. *James Robey.* 1872. *No.* 2,181.
   Chars sewage sludge from processes No. 1,954 of 1,868, or No. 1,354 of 1870, and uses it for treating sewage.

133. *Isaac Brown.* 1872. *No.* 2,279.
   A mere mechanical arrangement which aims at purifying sewage by settling, screening and straining.

134. *H. Y. D. Scott.* 1872. *No.* 2,538.
   Precipitates the sewage with lime, treats the effluent with the "phosphatic precipitants" mentioned in No. 849 of 1872, with or without charcoal. It is very difficult to find any essential difference between the process here described, and that given in No. 849 of 1872.

135. *B. W. Gerland and E. Johnson.* 1872. *No.* 2,569.
   Char fresh turf, and sawdust, spent tan, mixed with loam, and use the product for filtration, with or without the addition of phosphoric acid. They divide the sewage into two branches ; to the one is added phosphoric acid, with or without the charcoal dust, and to the other, milk of lime. The precipitate is let settle, and the effluent filtered through the charcoal abovementioned.

136. *W. Astrop.* 1872. *No.* 2,991.
   A mechanical method of separating the solids from the liquid portions of sewage.

137. *H. Y. D. Scott.* 1872. *No.* 3,028.
   A method of adding lime to sewage, and a process for drying sewage sludge.

138. *J. A. Manning.* 1872. *No.* 3,356.
   Evaporates sewage to dryness, passing the fumes given off into a furnace (!). The cost of this in the case of a large city?

139. *G. Alsing.* 1872. *No.* 3,412.
Converts sewage and night-soil into manure by mixing with sulphate of lime.

140. *C. Hills and B. Biggs.* 1872. *No.* 3,464.
Mix sewage in an air-tight tank with lime to liberate ammonia. Force air through the sewage into a second tank containing sulphurous acid. Or they force sulphurous acid into the sewage instead of air.

141. *D. Curran and James Dewar.* 1872. *No.* 3,533.
Use peat, either alone or along with chalk, lime, earth, etc., for filtering, disinfecting and absorbing foul waters.

142. *H. Y. D. Scott.* 1872. *No.* 3,755.
Uses the effluent from a lime-process for working water-closets.

143. *H. Y. D. Scott.* 1873. *No.* 154.
Treats sewage with lime in excess, and adds to the effluent "soluble salts of cheap metallic oxides." After this, the effluent may be filtered through charcoal, or may be run at once into a stream.

144. *J. L. D. Target.* 1873. *No.* 168.
Chars sewage solids mixed with sawdust, tar, etc., and uses them as fuel. Boils the sewage along with lime and catches the ammonia.

145. *E. C. Hamilton, W. R. Preston, and H. Jones.* 1873. *No.* 187.
Mix shoddy with sewage.

146. *James Robey.* 1873. *No.* 230.
Chars the sludge obtained under No. 1,954 of 1868, and No. 1,354 of 1870, with or without an admixture of clay, and uses it for treating sewage.

As the inventor claimed the use of the same sewage sludge charred, for the same purpose, in patent No. 2,181 of 1872, but abandoned the idea without proceeding to the great seal, the question might be raised, whether he did not, by so doing, anticipate the present patent?

147. *J. Jacobsen.* 1873. *No.* 266.

Treats sewage with phosphate of lime and sulphuric acid, diluted with the liquid sewage itself.

148. *H. Y. D. Scott.* 1873. *No.* 296.

Precipitates the suspended matters in sewage by limewater in excess. Treats the effluent with acid solution of phosphatic substances. If this is added in excess, the effluent is again treated with lime. Here we recognise a strong family likeness to the inventor's previous processes.

149. *Baldwin Latham.* 1873. *No.* 331.

Treats the deposits from various sewage processes—preferably that from No. 3,650 of 1869—with sulphuric acid to obtain fresh material for treating further portions of sewage.

The economy of this process, and of similar processes, is very doubtful, since a portion of sulphuric acid (*the* expensive article in making sulphate of alumina) is wasted by the organic matter and the lime present in the deposit.

150. *F. H. Atkins.* 1873. *No.* 556.

Filters sewage through ground coke or cinders converted by pressure into slabs. He also applies " galvanic, magnetic, or electric action to filtering apparatus, reservoirs, or tanks, for the purpose of precipitating organic and inorganic matters in suspension or solution."

151. *R. S. Symington.* 1873. *No.* 912.

An improvement on No. 2,667 of 1868. The effluent water is to be purified by "falling in a broken manner through a sufficient height before passing through the last filtering tank."

152. *G. Alsing.* 1873. *No.* 1,319.

Mixes sewage sludge with dry gypsum.

153. *H. Y. D. Scott.* 1873. *No.* 1,445.

The sewage deposit, of course from a lime process, is

treated with more lime, and mixed with some material wetted with sulphuric or hydrochloric acid. The compound obtained is used for deodorising pail-stuff, and finally employed as manure. Chloride of lime, or chloride of zinc or iron, or sulphates of those metals, may also be added to the compound.

Salts of zinc are injurious to vegetation, and cannot, therefore, be safely added to any kind of manure.

154. *H. Y. D. Scott.* 1873. *No.* 1,509.

Dries sewage deposits from lime processes; are dried in retorts and used along with lime, chloride of lime and charcoal.

155. *Walter Brown.* 1873. *No.* 1,555.

Calcines shaly minerals with exclusion of air, quenches them with water, and uses the pieces for making a filter-bed, or for treating sewage in other ways.

It must be remembered that many "shaly minerals" contain nothing capable of yielding a soluble salt of alumina after ignition.

156. *E. Moriarty.* 1873. *No.* 1,686.

Treats sewage with, per 28 lbs. :—12 ozs. acid phosphate of magnesia, 6 ozs. sulphate of iron, 4 ozs. sulphur (!), 4 ozs. ammonia, 4 ozs. phosphoric acid, 4 ozs. nitrate of soda, 3 ozs. nitrate of potash, 4 lbs. gas tar, 6 lbs. wood charcoal, and sulphate of lime to bring the whole to a solid state.

This process does not aim at producing an effluent, and is intended for cesspool matters rather than for town sewage.

157. *B. Green.* 1873. *No.* 1,885.

Draws sewage gases through a fire, lets solids deposit as manure, and runs the liquid into a river.

158. *Jos. Townsend.* 1873. *No.* 1,967.

Treats sewage with any of the three following mixtures :—1. 100 lbs. of a phosphate containing 40 per cent.

phosphoric acid and 20 per cent. alumina is mixed with 50 lbs. of lime, and by preference 2 to 5 per cent. of soda or potash, or an equivalent quantity of carbonate, sulphate, or sulphite of soda or potash, " with sufficient lime to set free the alkali." The resulting products are " principally phosphate and aluminate of lime." For the lime may be substituted 36 lbs. magnesia, or 47 lbs. lime and 8 lbs. magnesia. 2. A mixture of alumina with lime or magnesia, or both lime and magnesia. 3. A "substance containing alumina associated with silica" is mixed with lime or magnesia and alkali, or "substances yielding alkali" are added.

159. *John Leigh.* 1873. *No.* 2,071.

Adds a solution of an earthy salt, followed by a solution of silicate of soda or potash. As earthy salt chloride of lime may be used (!). If much gelatinous or aluminous matter is present tannin is also added.

For "aluminous" the correct reading is probably "albuminous."

160. *Jeremiah Marsden and J. Collins.* 1873. *No.* 2,317.

Treat sewage per 200,000 gallons with 12 cwt. lime, 40 cwt. coal ashes, 16 cwt. charcoal, "and a small quantity of an acid salt of soda, potash, iron, manganese, or the like."

This process is, or has been, at use at Bolton. How manganese can be called "the like" of soda or potash is not apparent. Coal ashes are of very doubtful value.

161. *Robert Knott.* 1873. *No.* 2,442.

Treats sewage with a mixture of quicklime and soda introduced into the sewage in a boiling state.

162. *F. Jacobsen.* 1873. *No.* 2,454.

Precipitates sewage with the refuse "obtained after the lye-water of paper-mills has undergone the soda-recovering process." To facilitate precipitation he further adds common salt, sulphate of zinc, chloride of iron, alum, slaked lime, and "the water from electric batteries."

Sulphate of zinc is fatal, common salt useless, and the water from batteries very doubtful.

163. *F. Jacobsen.* 1873. *No.* 2,455.

Treats the waste water from paper-mills and other works with lime. To assist the process he adds "common salt, sulphate of zinc, chloride of iron, and perchloride of iron."

164. *William White.* 1873. *No.* 2,532.

Neutralises sewage with lime if necessary, and treats with sufficient chloride of calcium to precipitate sulphates, carbonates, and phosphates along with albuminous and other matters. The solution is then treated with sulphate of iron to convert the excess of chloride of calcium into sulphate, setting free chloride of iron, which may be precipitated as oxide by the addition of lime.

The final effluent will therefore be alkaline after these four successive treatments. In the deposit there will be sulphate of lime, which, as already explained, is not to be desired.

165. *James Robey.* 1873. *No.* 2,534.

Treats sewage with raw peat, and adds any suitable precipitating agent.

166. *C. Rawson, W. C. Sillar, J. W. Slater, and T. S. Wilson.* 1873. *No.* 2,662.

Pass sulphurous acid, chlorine, or other disinfecting gases through sewage, and then precipitate with sulphate of alumina, etc., if needful.

This process was mainly intended for blood. For sewage it is too costly and circumstantial.

167. *R. Goodall.* 1873. *No.* 2,791.

Adds per 1,000 gallons 2 bushels of fine ashes and

23 lbs. slaked lime. After agitation, 1 pint sulphate of iron and 7 pints of solution of sulphate of magnesia.

168. *W. Whitthread.* 1873. *No.* 3,169.

Treats sewage with manganese tetra-chloride. Other haloid salts of manganese may be used. Finally, an alkali is added.

169. *A. C. Fraser and W. Watson.* 1873. *No.* 3,632.

Treat sewage with calcined schist, which may be saturated with dilute sulphuric acid, and mixed with clay and sulphate of lime. The outflow from the last tank passes into a filter charged with lime, calcined schist, charcoal and sand.

170. *H. Y. D. Scott.* 1873. *No.* 3,742.

Sewage is precipitated in the "ordinary" manner with lime, and the effluent is treated with "an impure sulphate or chloride of lime and alumina," prepared by "digesting prussiate of potash charcoal with dilute sulphuric or hydrochloric acid."

Chloride of lime cannot be obtained at all in the way described. Alumina is present in the charcoal merely as an impurity present in the carbonate of potash used in prussiate-making. The product really obtained will consist mainly of sulphate or chloride of iron, which can be more conveniently or cheaply obtained in other manners.

171. *W. White.* 1873. *No.* 3,781.

Precipitates sewage with lime-charcoal, a calcined mixture of chalk, peat, and sawdust. Soluble phosphate may be added to the sewage before the treatment with lime-charcoal. "Lime-coke" may also be used, made by burning a mixture of small coal and lime. Ground mineral phosphates may be mixed with peat, or other form of carbon, and used as above. Industrial waste waters unfit for manure may be treated with "lime-clay-charcoal," a calcined mixture of lime and clay with or without carbon.

172. *W. White.* 1873. *No.* 3,781.

Disclaims the method of preparing the precipitating agents mentioned in the last specification.

173. *B. Green.* 1873. *No.* 3,833.

Lets the sewage settle in pits, carrying the gases into a furnace, and letting the liquid run into "a river, the sea, or elsewhere."

This patent, like certain others, is based on the mistaken notion that sewage, after spontaneously depositing its "sludge," is harmless, and may be safely run off into rivers.

174. *Paul Curie.* 1873. *No.* 4,181.

Treats the sewage with a "disinfectant," sulphate of iron, clay, chalk, etc. The products of combustion from a furnace, mixed with air, are driven through the sewage, which is then apparently evaporated to dryness.

175. *A. E. Schmersahl.* 1874. *No.* 160.

Treats sewage with a mixture of 2 parts sulphuric acid and 1 part hydrochloric acid, and then with lime.

The result of such treatment will be the formation of sulphate of lime—an evil—and chloride of calcium.

176. *H. M. Synge.* 1874. *No.* 255.

A complicated filtration process through successive tanks charged with materials of increasing fineness.

177. *E. H. C. Monckton.* 1874. *No.* 265.

Purifies sewage by passage through electrified channels, or drives ozonised water into sewage for the purpose of purifying it. Recovers metals in solution from the sewage of manufacturing towns by an electric process. Uses windmills as a power to generate electricity for purifying sewage.

Ozone would doubtless prove a powerful deodorising

agent for sewage if its price permitted. It might be interesting to try the action of electricity on sewage where power for driving dynamos could be had free of cost.

178. *H. Y. D. Scott.* 1874. *No.* 283.

Precipitates sewage by lime, or removes the solids by other means. Passes liquid through filters charged with " phosphate of iron salts," or adds such salts to the liquid. Runs effluent through filters of lime and charcoal. Gives methods for making the phosphate of iron salts, to which, before use, lime refuse from soda or gasworks may be added.

The effluent here will apparently be alkaline. It is interesting to note how many changes, capable of being patented, have been rung on the old lime process, in which its cardinal faults are retained, whilst its cheapness and simplicity are lost.

179. *H. Y. D. Scott.* 1874. *No.* 653.

Precipitates sewage by lime mixed with soluble phosphate, or phosphoric acid, or soluble metallic salts.

Lime mixed with " soluble metallic salts " will precipitate those most likely to be used for sewage purposes, and render them comparatively inert. It will also render the soluble phosphate or the phosphoric acid insoluble. Phosphoric acid is, further, an agent far too costly to be used in sewage treatment.

180. *Rupert Goodall.* 1874. *No.* 848.

The author places in one vessel a " mixture of calcium and carbon," in another slaked lime and carbon, or slaked lime alone, and in a third vessel a solution of sesqui-persulphate of iron, and delivers these into the sewage.

181. *W. R. Lake.* 1874. *No.* 1,415.

Boils the sewage, and receives the volatile matters in a series of Woolf's bottles containing sulphate of iron, lime, sulphuric acid, etc., to fix the ammonia. The residue left in the boilers is cooled and used as manure.

How the residue is to be rendered inoffensive we are not informed. *Quaere*—How many boilers would be required to receive the sewage of London ?

182. *J. Towle.* 1874. *No.* 1,426.

Delivers sewage into pits into which town refuse is also cast. Spent hops and cut straw are also placed in the pits.

183. *J. H. Kidd.* 1874. *No.* 1,764.

Allows the sewage solids to settle in tanks and runs off the effluent water. Adds, apparently to the deposit, salt or lime and carbonised shale.

184. *Rupert Goodall.* 1874. *No.* 1,826.

Treats sewage per 1,000 gallons with 1 quart of sesquipersulphate of iron, and 1 pint thereof, and 1 pint of nitrate of iron. There may also be added along with the iron liquors saturated solutions of lead salts, or of ferro-cyanide of potassium.

Or a mixture of gas-lime 1 or 2 parts, slaked lime 1 or 2 parts, and animal carbon 2 to 4 parts may be added in the proportion of 27 lbs. per 1,000 gallons. After agitation, iron salts as above are added, or a mixture of 10 parts animal carbon, 2 or 3 brown oil of vitriol, or 3 or 4 nitrosulphuric acid, after which the whole is again agitated.

This patent, it is believed, was assigned to the Rivers Clarification Company, Limited, of Leeds.

185. *W. H. Hughan.* 1874. *No.* 1,959.

Treats the sewage with an "antiseptic" made of Portland cement, sulphates of soda, magnesia and potash mixed in oil, preferably mineral oil, and then precipitates with a mixture of Portland cement, fluorspar and oil. The effluent is filtered, and the precipitate with the residue from the filtration is mixed with hot superphosphate. The "antiseptic" may also be made from seaweed, clay and soda-waste treated with sulphuric acid.

186. *S. Hallsworth and R. Bailes.* 1874. *No.* 2,408.

Agitate sewage in a tank with either persulphate of iron made from iron pyrites, or a mixed solution of sulphuric acid and iron from the beds of coal and iron pyrites, or sulphuric acid mixed with the mother-liquor left after crystallisation of copperas, or a mixture of sulphuric acid with other equivalent or suitable solution of iron, or of a solution of copperas and sulphuric acid.

187. *A. E. Schmersahl.* 1874. *No.* 2,439.

Treats sewage with a mixture of 2 parts dilute sulphuric acid and 1 part hydrochloric acid, enough to make the sewage acid. Or the acids may be added separately, or phospho-muriate of lime, commonly called bone-liquor, may be employed, or chloride of manganese. Sufficient milk of lime is then added to neutralise the acid. See the inventor's previous patent (No 160, of 1874).

188. *W. A. Lyttle.* 1874. *No.* 2,446.

Gives directions for deodorising sludge; how it is to be precipitated the inventor does not state.

189. *H. Y. D. Scott.* 1874. *No.* 2,450.

The object of this patent is the recovery of carbonate of ammonia from sewage. If a manure is to be made, magnesia and phosphate of magnesia are agitated with the liquid to be treated.

190. *W. Spence.* 1874. *No.* 2,461.

Treats sewage in lead-lined tanks having false bottoms of copper. "A coil of pipes is inserted in such a manner that the half of the pipes are above and the other half below the false bottom." To the first tank acid is added in quantity sufficient to fix the ammonia and dissolve the phosphates of lime. Steam is then admitted into the coils.

There is no mention of any phosphate of lime having been added.

191. *G. Willett, R. J. Harris, and James Lund.* 1874. *No.* 2,567.

Filter through tanks of coal-ashes or coke.

192. *H. Y. D. Scott.* 1874. *No.* 2,568.

Treats phosphates with acids, adds magnesia, and uses the product to fix the ammonia of urinous or other ammoniacal liquids.

193. *J. H. Kidd.* 1874. *No.* 3,199.

Lets sewage settle, runs off effluent and dries the solids.

194. *James McIntyre.* 1874. *No.* 3,225.

Depositing tanks without the use of any precipitant or deodorant.

195. *S. Hallsworth and R. Bailes.* 1874. *No.* 3,459.

The inventors take per 5,000 gallons of sewage, 1½ gallons of pyrites liquor and 50 lbs. of slaked lime, or 100 lbs. of gas lime.

196. *A. F. Paget.* 1874. *No.* 3,613.

Treats waste waters with a mixture of sesqui-chloride of iron and aluminium chloride, and afterwards with lime-water. The chloride solution is obtained by dissolving any argillaceous iron ore containing phosphoric acid in hydrochloric acid. Carbolic acid, permanganate, or other disinfectants may be added as the water leaves the mixing tanks.

197. *G. Mackay.* 1874. *No.* 3,751.

Treats waste waters with per-salts of iron, preferably the perchloride or persulphate, and then with lime or other suitable alkali.

198. *V. B. Halle.* 1874. *No.* 3,784.

A mere process of precipitation with lime followed by filtration.

199. *F. T. Bond.* 1874. *No.* 3,799.

Uses for disinfection sulphates of iron, aluminium and copper, carbolic acid, terebene potassium, bichromate and permanganate.

200. *Rupert Goodall.* 1874. *No.* 4,158.

Mixes ashes and gas lime or waste lime from ammonia works with sulphuric acid till effervescence ceases ; adds water and uses one to two gallons of this to 1,000 gallons of sewage.

The sewage is first treated with slaked lime, or with a mixture of six parts slaked lime and one part animal carbon. Five pounds of lime or 20 lbs. of the compound may be used for every 1,000 gallons of sewage and ½ to 1 gallon of the first described mixture is then stirred in.

201. *J. C. Morrell.* 1874. *No.* 4,247.

Treats sewage with alum, quicklime, or a crystallised chemical substance, which does not appear to be named or described.

202. *H. Y. D. Scott.* 1874. *No.* 4,305.

Lets sewage deposit in a tank. To the effluent from this he adds milk of lime in a second tank. The lime effluent is then treated in another tank with an acid solution of phosphate of iron, lime, and alumina.

Successive precipitations with different agents are bad on account of the great outlay for plant and labour which they involve.

203. *W. J. Pughsley.* 1,874. *No.* 4,373.

Treats refuse liquor from tin-plate works by filtering first through limestone, then through charcoal, and then through bone-ash. The two latter are separated from each other by a perforated board, and the bones are supported upon an iron plate. The bones are removed from time to time and used as manure.

204. *G. Mackay.* 1875. *No.* 91.

Treats sewage with mixed solutions of perchloride and persulphate of iron. Salts obtained from alum, alkali, and galvanising works may be mixed with the solution. Lime or other alkali is then added.

205. *G. Rydill.* 1875. *No.* 150.

Filters through a bed of ashes.

206. *J. Box, E. Aubertin, L. Boblique, and H. Leplay.* 1875. *No.* 214.

Treat sewage first with a ferruginous phosphate of soda or potash and then with a salt of magnesium.

207. *G. Rydill.* 1875. *No.* 399.

Treats sewage, etc., with caustic soda or lime. Filters through ashes, and forces air through it from perforated pipes.

208. *J. Hallsworth and R. Bailes.* 1875. *No.* 573.

Treat sewage with two clarifying mixtures. 1. Spent residues of iron pyrites or other ores containing iron as peroxide are ground and mixed with an equal weight of any of the following : Copperas, dry copperas, copperas bottoms, copperas sediment, or sediment from the manufacture of nitrate of iron. The mixture is calcined and allowed to cool. 2. The pyrites residue or iron ore is saturated with liquor from the pyrite beds at copperas works, or with dilute sulphuric or hydrochloric acid. The sewage is first treated with slaked lime or with calcium (?) in the proportion of 78 lbs. of the former or 156 lbs. of the latter to 5,000 gallons of liquid, and mixed together ; 23 lbs. of mixture No. 1 or 46 lbs. of No. 2 is then mixed with a portion of sewage and then added to the lime sewage.

209. *W. M. Brown.* 1875. *No.* 1,335.

A kind of filtration process.

210. *T. Page.* 1875. *No.* 1,625.

A screening arrangement, the deodorising agent is named.

211. *P. Spence and F. M. Spence.* 1875. *No.* 1,704.

Manufacture of alumino-ferric cake, used in sewage treatment.

212. *J. Hill.* 1875. *No.* 1,745.

A contrivance for expressing moisture which the inventor applies to separating the solid from the liquid portions of sewage.

213. *John Yule.* 1875. *No.* 1,759.

A barge for conveying sewage.

214. *M. F. Anderson.* 1875. *No.* 1,845.

Treats sewage sludge with coprolite, phosphorite, or ground bone, together with sulphuric acid.

215. *J. J. Coleman.* 1875. *No.* 1,954.

Filters sewage through spent shale from the mineral oil works, or the shale may be added in the sewers and the sewage allowed to settle in tanks. The effluent is passed through beds of shale. See Specification of R. Irvine, 1886, No. 2,218.

216. *P. Spence and F. M. Spence.* 1875. *No.* 1,961.

Improvements in alumino-ferric cake, rendering it more basic and better adapted for treating sewage.

217. *D. Wilks.* 1875. *No.* 1,972.

Treats sewage in tanks with powder obtained by carbonising town refuse. The effluent is filtered.

218. *J. Odams and R. Blackburn.* 1875. *No.* 2,358.

Revolving screens for separating the solids from sewage.

219. *J. Hanson.* 1875. *No.* 2,675.

Precipitates sewage with, per 100,000 gallons, slaked lime 20 to 23 lbs., soot or flue dust $\frac{1}{2}$ lb., black-ash (by which the inventor means oat waste from the alkali works) 30 lbs.

220. *T. Stevens.* 1875. *No.* 2,829.

Adds sulphate of lime and common salt to the sewage on its way to the settling tank. Adds milk of lime as it enters the tank, and finally filters.

221. *A. M. Clark.* 1875. *No.* 3,162.

Treats sewage, according to circumstances, with—1. Aluminate of soda, alone or in conjunction with soluble alkaline or earthy phosphates. 2. The same phosphates mixed or combined with oxide or phosphate of iron along with a soluble salt of magnesia. 3. Oxides of nitrogen higher than nitrous oxide but lower than nitric acid.

In consequence of this patent, no general claim to the use of aluminate of soda can be maintained by subsequent inventors.

222. *J. W. Slater.* 1875. *No.* 3,368.

Ignites sewage sludge in retorts, obtaining a carbon which may be used in treating gas. The volatile products are condensed and yield ammonia and illuminating gas.

The gas thus obtained is of " 8 candle " power.

223. *C. Rawson and J. W. Slater.* 1875. *No.* 3,703.

Use in place of alum aluminous shales, such as those of Campsie, either in their raw state or artificially weathered by treatment with super-heated steam, or steam and atmospheric air, or hot air charged with moisture, or sulphurous acid gas along with air or steam.

224. *A. Le Tellier.* 1875. *No.* 4,061.

A mechanical arrangement for treating sewage.

225. *James Bannehr.* 1875. *No.* 4,122.

A mechanical arrangement. The effluent is filtered and treated with electric currents.

226. *H. M. Ramsay.* 1875. *No.* 4,420.

Filters sewage first over loose charcoal and scrap metals,

and then over "porous-plastic carbon." These filters may be cleansed by reburning or by forcing clean water through them.

Where, in the latter case, is this water to go?

227. *John Hanson.* 1876. *No.* 225.

Improvement on No. 2,675, of 1875. Blast-furnace slag and "Paris white" (*i.e.*, whiting), with or without the addition of sulphuric or hydro-chloric acid, are used to precipitate sewage, either alone or mixed with the ingredients named in the former specification.

If the slag is added without acid, it is inert. If acid is used it will be neutralised by the "Paris white."

228. *H. Y. D. Scott.* 1876. *No.* 322.

A complicated process for treating solid excreta and pail-stuff.

229. *F. Hille.* 1876. *No.* 1,355.

Slakes lime with sea-water, or with solutions of magnesian salts, or with the refuse liquor from salt works, dried, calcined and re-dissolved in water, with 5 lbs. gas tar to every 100 lbs. of lime. This paste is mixed with water, and applied to sewage in tanks. The effluent from this process may be treated with iron-perchloride or with carbonic acid gas.

230. *J. Bannehr and S. A. Varley.* 1876. *No.* 1,739.

A mechanical arrangement for intercepting solids. Air is then forced through the liquid; such air may be ozonised by means of electric currents.

231. *C. Rawson and J. W. Slater.* 1876. *No.* 1,893.

Treat sewage with double fluorides and silico-fluorides. Make an artifical animal charcoal by treating phosphates with carboniferous and nitrogenous matters.

232. *W. Clark.* 1876. *No.* 1,930.

Mixes sewage with an acid to fix ammonia, draws off clear liquid and evaporates. Or treats sewage with lime, and drives off the ammonia in a "triple effect evaporator." The ammonia is condensed, and the sludge filtered and pressed.

233 *G. Bischof.* 1876. *No.* 2,080.

Purifies sewage either by means of spongy iron or manganic dioxide, employed either as filtering media, or added in powder to the sewage.

234. *W. Webb.* 1876. *No.* 2,124.

A mechanical arrangement for separating the liquid from the solid portions of sewage.

235. *Thomas Lovell.* 1876. *No.* 2,387.

Merely arrangements for irrigation.

236. *F. W. F. Reinhold Goedicke.* 1876. *No.* 2,526.

Separation of solids from liquids, the latter being applied for irrigation or other purposes.

237. *J. W. Slater.* 1876. *No.* 3,095.

Treats sewage with alkaline, alkaline earthy, earthy or metallic salts of hypo-sulphurous, sulphurous and thionic acids. These are obtained by exposing tank-waste to air. They may be used along with aluminium sulphate or chloride. Claims also infusorial earth in combination with salts of alumina, hydrated silica, and clay, or as a filter bed.

238. *James Miller.* 1876. *No.* 3,107.

Passes sewage through tanks with perforated sides filled alternately with gravel and sand, or lime and sand.

239. *Henry Staples.* 1876. *No.* 3,307.

Treats natural sulphates of alumina, Campsie shale, etc., with hydrochloric acid in heat, and uses the product for treating sewage. Adds to the compound alkali waste to take up moisture, but not to decompose salts of alumina or iron, or to give the mixture on alkaline reaction.

240. *Joe Frost.* 1876. *No.* 3,365.

Precipitates sewage with sulphides and hydrate of

barium, hydrates, precipitated hydrates, sulphydrates and sulphides of the same metal, sulphides of soda, slaked lime, carbonate of lime and sulphuric acid.

The poisonous character of the salts of barium is well known.

241. *W. White.* 1876. *No.* 3,576.

Treats sewage with oxychloride or chloroxide of calcium, obtained by boiling lime in a strong solution of calcium chloride, or quicklime may be slaked with dilute hydrochloric acid. Sulphates of aluminium, iron or zinc may be used in connection, as also charcoal and chloride of lime.

The poisonous nature of the sulphate of zinc, and the inadmissibility of chloride of lime have been already explained.

242. *A. Greenwood, G. E. Davis, and J. J. Speakman.* 1876. *No.* 3,673.

Place liquid sewage in a tank, and allow it to remain until the nitrogen has been transformed into ammonia (a great nuisance!), which is then driven off and fixed.

243. *J. Watson.* 1876. *No.* 4,203.

Treats sewage with hydrochloric acid in a tank, and uses the precipitated sludge for manure.

244. *C. D. Abel.* 1846. *No.* 4,516.

A complicated system of settling and filtration.

245. *G. Rydill.* 1876. *No.* 4,848.

Precipitates sewage with waste liquors from the treatment of woollen, silk, and other manufactures, also waste acid liquors which have been used for separating animal and vegetable fibres. Treats sewage in a filter-bed made of "extracted vegetable substances, dust and foreign matter," or "vegetable substances containing acid; chemicals are used as a disinfectant, with earth alum, shale, clay refuse, animal carbon, ashes, lime, or salt, the same being strongly impregnated with sulphuric, hydrochloric or nitric acid, or alum along with clay or other chemical agents."

In other words, the inventor claims things in general, useful or useless, compatible or incompatible.

246. *H. D. Y. Scott.* 1877. *No.* 103.

Treats pail-stuff and gas liquor for manures.

247. *Houzeau and others.* 1877. *No.* 263.

Use for sewage, coal-ashes of all kinds from dwellings and manufactories; pyritous, ligneous, or sulphurous ashes, sulphatized or not; ashes and residues from manufactories of soda, sulphuric acid, sulphate of iron, and waste products from polishing glass; all refuse of products having been used for industrial purposes (!); gas from the crystallization of alums, sulphate of alumina, sulphate of iron, and all crystalline salts. These bodies are lixiviated, with or without the aid of an acid, and added to the sewage along with (in most cases) milk of lime.

This is, perhaps, the most singular specification ever drawn up as far as sewage treatment is concerned. Among the substances claimed under the head "All refuse of products having been used for industrial purposes," there are not a few totally unsuitable, and others which have been used and claimed before. Among the former may be mentioned the highly poisonous refuse of the manufacture of magenta, and among the latter a variety of aluminous products. French inventors seem, as a rule, utterly to ignore the fact that, according to English law, a bad claim imperils the entire patent, and they re-invent with great calmness processes which are already public property.

248. *J. H. Johnson (Emile Barrault).* 1877. *No.* 488.

Claims manufacture of sulphate of alumina by the action of sulphuric acid upon dried or powdered shales, and uses the sulphate thus obtained in treating sewage (! !).

249. *G. Lunge.* 1877. *No.* 638.

Purifies the drainage from alkali waste.

250. *R. Turnbull.* 1877. *No.* 824.

Mechanical arrangements only.

251. *G. Alsing.* 1877. *No.* 835.

Filters through a screen, and treats with a solution of hydrate and sulphide of calcium and gibbside (*sic*) and filters effluent again.

252. *John Hanson.* 1877. *No.* 860.

Treats sewage with waste haematite, sulphurous gas, a mixture of alum, soda, and " black ash refuse," *i.e.*, tank waste.

Softens water with carbonate of potash and sometimes nitre-cake.

253. *S. Hallsworth and R. Bailes.* 1877. *No.* 952.

Treat sewage with (1) 1,000 gallons of liquor from beds of iron pyrites (50° Tw.), mixed with 250 lbs. sulphuric acid at 144° Tw. (2) 1,000 gallons of dissolved copperas, or copperas bottoms, or other liquor from copperas at 40° Tw. with 370 lbs. sulphuric acid at 144° Tw. (3) 1,000 lbs. solution of iron at 40° Tw. from pickling iron at wire and sheet iron works, along with 250 lbs. sulphuric acid at 144° Tw. (4) A liquor from iron byrites and sulphuric acid and water neutralized with scrap iron, set at 75° Twaddle, adding to each 1,000 gallons 250 lbs. sulphuric acid at 154° Tw. . . . (6) tapcinder from puddling-furnaces, or other slags containing protoxide of iron and dissolved in sulphuric acid. (7) nitrate of iron at 4° Tw. To 10,000 gallons of the sewage they add any of the aforesaid mixtures, followed, or, in some cases, preceded, by milk of lime.

This specification might be advantageously studied by inventors who propose copperas and lime as a novelty in sewage-treatment.

254. *M. H. Singe. No.* 1,619.
Filtering and purifying apparatus.
255. *H. Y. D. Scott. No.* 1,772.
A magnesian lime-process.
256. *J. Fenton. No.* 1,982.
A filtering arrangement.
257. *W. R. Lake. No.* 2,661.
Prepares an artificial bone black.
258. *J. Hanson. No.* 2,725.

Treats sewage with six liquids, viz., muriate or nitro-muriate of tin at 50° Tw.; silicate of soda dissolved in 7 parts of water; borax dissolved in 9 parts of water; "litchen" or Iceland moss boiled in water; a mixture of the tin solution and of the decoction of Iceland moss, and peracetate of iron in 4 parts of water. In conjunction with the above may be used carbonates or hydrated oxides of iron deposited by water draining from mines and dissolved in sulphuric acid, asbestos ground in water, slaked lime, tank-waste, and sulphurous acid gas.

Salts of tin are doubtless good precipitating agents, but they are too costly and too poisonous. Sulphurous acid gas has been already noticed.

259. *C. J. Wollaston. No.* 2,841.
Uses sulphurous acid or chlorine, or both (!) in conjunction with lime and magnesia.

The fact that the actions of sulphurous acid and of chlorine are antagonistic, the former being a de-oxidising and the latter an oxidising agent, is here overlooked.

260. *B. B. Standen.* 1877. *No.* 3,395.
   No agents for treating sewage are here proposed.

261. *H. Y. D. Scott.* 1877. *No.* 3,977.
   Proposes no means of treating sewage.

262. *J. Gray.* 1877. *No.* 3,571.
   Merely a lime process. The lime is slaked with boiling water, or alum water, and is then used for treating. The alum is, of course, decomposed!

263. *H. Robinson and J. C. Melliss.* 1878. *No.* 12.
   Treat sewage with the joint addition of copperas and sulphate of alumina, or, if necessary, lime. *Quaere*, the novelty?

264. *J. Foulis and J. A. Carrick.* 1878. *No.* 73.
   Treat waste waters with sulphate or chloride of zinc, waste bleach, and soda-limes (?) separately or in conjunction.

265. *Walter East.* 1878. *No.* 92.
   Adds putrescent sewage to hasten fermentation. Conveys away noxious gases in pipes. Adds iron oxide to remove sulphuretted hydrogen.

266. *R. U. Etzensberger.* 1878. *No.* 264.
   Does not bear upon sewage.

267. *J. Adamson and H. Booth.* 1878. *No.* 2,937.
   A lime process.

268. *W. Pochin.* 1878. *No.* 4,270.
   Treats iron slag with sulphuric acid. No mention of sewage.

269. *W. R. Lake.* 1879. *No.* 179.
   Enriching phosphate of lime. No reference to sewage.

270. *W. H. Denham.* 1879. *No.* 437.
   Burns household refuse and precipitates sewage with the ashes.

271. *H. Chamberlain.* 1879. *No.* 2,599.
A straining process.
272. 1879. *No.* 2,345.
Irrelevant.
273. *W. L. Wise (Loewig).* 1879. *No.* 3,195.
Prepares "carbonic alkali of alumina."
274. *T. H. Cobley.* 1879. *No.* 3,312.
No reference to sewage.
275. *R. Wild.* 1879. *No.* 3,373.
Treats with lime and alum and "other suitable precipitant," and filters.
276. *R. Wild and H. Ledger.* 1879. *No.* 3,980.
Mechanical arrangements.
277. *W. F. Mast.* 1879. *No.* 4,402.
Method of extracting ammonia from excrement and urine.
278. *H. C. Bull.* 1879. *No.* 5,324.
No reference to sewage.
279. *J. C. Mewbari.* 1880. *No.* 243.
Purifies wool-washings with caustic lime, epsoms and copperas, forming insoluble soaps.
280. *J. G. Tongue.* 1880. *No.* 742.
Destroys organic matter in waste waters by heating them and applying lime white, chloride of magnesium, and other "suitable chemicals." Special mechanical arrangements.
281. *J. Duke.* 1880. *No.* 748.
Filters through silicates of lime, potash, soda or alumina separately or combined, peat charcoal, and superphosphates.
282. *C. Dickinson.* 1880. *No.* 3,898.
Proposes mechanical arrangements. Uses for sewage treatment salt-cake, alum and potash, and, if necessary, "well-known deodorising means."

Salt-cake is, of course, useless, and potash, if it answers any good end, is very expensive.

283. *E. Parry and T. H. Cobley.* 1880. *No.* 3,554.

Prepare earthy silicates, and make no reference to sewage.

284. *J. Duke.* 1880. *No.* 2,994.

Manufactures soluble silicates by treating Roman and Portland cement, or suitable natural silicates, or kainite, with sulphuric or hydrochloric acid. The resulting product is applied to manures, but not to the treatment of sewage.

The part which kainite can play in the preparation of silicates is very questionable.

285. *J. H. Johnson.* 1880. *No.* 4,603.

Treats lavas with sulphuric acid or hydrochloric acid, and uses the resulting mixture, "lava-syrup," as a disinfectant or for treating sewage. Uses also alumnite calcined in presence of potassium chloride.

286. *W. H. Denham.* 1881. *No.* 1,413.

Mechanical arrangements.

287. *W. R. Lake.* 1881. *No.* 1,448.

A process for recovering muriatic acid used in treating bones. No reference to sewage.

288. *R. Wild and H. Ledger.* 1881. *No.* 1,564.

Claim treatment of sewage with alumino-ferric cake and soap water, or in combination with ammonia. Twofold precipitation.

Soap water is one of the impurities which have to be removed from sewage. The addition of ammonia to sewage is an unhappy idea. Alumino-ferric cake has been used for sewage treatment ever since it appeared in commerce.

289. *J. Storr.* 1881. *No.* 1,716.

Proposes a process for obtaining ammonia from sewage, etc. Provisional protection was not granted.

290. *H. Collett.* 1881. *No.* 2,004.

Treats sewage with "vitriolic powder," which is, as he says, "a special substance prepared from pyrites," and with or without the addition of crude zinc sulphate and sulphuric acid. Further, he adds, or may add, to the "separated liquid"—probably to the effluent water from this first precipitation—chalk, limestone or other matter containing carbonate of lime, slaked lime, carbonate or sulphate of soda. Instead of "vitriolic powder" may be used silicates of soda or potash, zinc or iron sulphates, fluorides of silicon or boron, or hydrofluosilicic acid, bark, sumac, peat, charcoal powder.

In this strange specification the preparation of " vitriolic powder" is nowhere described—a serious omission, since there is no article known in commerce under that name. Salts of zinc are, from their poisonous nature, quite out of the question. Sulphate of soda is useless, and carbonate of soda is not only too costly, but injurious. I do not suppose that this process has ever been worked upon a practical scale.

291. *Baron Adelbert von Podewils.* 1881. *No.* 2,295.

Proposes arrangements for disinfecting fæcal matters by fumigation.

292. *W. R. Lake (F. Petri).* 1881. *No.* 2,345.

Uses as a disinfectant peat-waste, etc., moistened with carbolic acid and chloroform ; a powdered mixture of copperas and coke, with a solution of nitro-benzol in alcohol. To stale fæcal matters he adds chloride of lime and alcohol ! ! !

This most extraordinary process is apparently intended not so much for sewage as for night-soil, pail-stuff, the contents of cess-pools, etc.

293. *G. W. von Nawrocki.* 1881. *No.* 2,790.

Proposes an arrangement for evaporating and drying sewage.

294. H. E. Newton. 1881. *No.* 4,088.

Adds to sewage first an astringent, then milk of lime, sulphates of iron or manganese, or powdered gypsum. The fæcal matters are then distilled with alkali so as to liberate ammonia.

295. A. M. Clark. 1881. *No.* 5,199.

Treats sewage with chlorides and fluorides, *e.g.*, chloride of zinc, potash and phosphate of lime, or chloride of manganese potash, silicate of potash and alumina in the form of phosphate or carbonate.

Chloride of manganese was an excellent precipitant, but it is no longer obtainable as a waste product. Chloride of zinc from its poisonous nature is inadmissible. Potash and silicate of potash are much dearer than soda and silicate of soda, without having any advantages. Carbonate of alumina is a body whose existence is doubtful. Carbonates as a class are injurious in sewage precipitation, since the carbonic gas which they give off in contact with acids or acid salts stirs up the deposit.

296. F. Petri. 1881. *No.* 5,390.

Proposes a filtration process. Disinfects with carbolic or salicylic acid, ethylic or triethylic chlorides, or other chlorides of alcohol.

297. Peter Lowe. 1882. *No.* 268.

Proposes arrangements for straining sewage.

298. H. Y. D. Scott. 1882. *No.* 311.

A process for the manufacture of manure.

299. John Brock. 1882. *No.* 473.

Manufactures manure from alkali waste !

300. G. J. Andrews and F. H. Parker. 1882. *No.* 1,396.

Treat sewage with muriatic acid and " alkaline soda," They then add sulphate of iron.

What " alkaline soda " may be can only be guessed.

As chlorine gas is said to be evolved by the meeting of these chemicals, it must probably be chloride of soda, otherwise known as bleaching soda, or sodium hypochlorite, which is not a very cheap article.

301. *E. Edwards.* 1882. *No.* 2,970.

Proposes a scheme for treating sewage in closed receptacles.

302. *J. C. W. Stanley.* 1882. *No.* 3,091.

Makes paints from river mud. There is no mention of the purification of sewage.

303. *T. H. Cobley.* 1882. *No.* 3,497.

Treats sewage with chlorides of aluminium, calcium, iron and magnesium, alone or in combination with carbonaceous matter. These substances are principally obtained from basic clay, cinder and scoriæ, "shale-ash clay" and bauxite, treated with hydrochloride acid.

"Shale ash clay" is a doubtful substance. The abovementioned chlorides have all been previously claimed or mentioned in patents for the treatment of sewage.

304. *James Young.* 1882. *No.* 3,562.

Treats sewage by distillation, with or without the addition of lime, and collects the ammonia given off.

305. *R. Nicholls.* 1882. *No.* 3,863.

Proposes a scheme for enabling each house to treat its own excreta. No novel chemicals are claimed.

306. *James Young.* 1882. *No.* 4,659.

Expels ammonia from sewage by passing into it steam or steam and air. The sewage passes through a tower in one direction, whilst the steam and air travel in the opposite direction.

307. *S. Walter.* 1882. *No.* 5,285.

Treats sewage with waste lime and lime-water, and filters.

Waste lime is not a very abundant article, and brine of any kind is an undesirable addition to sewage.

308. *F. Petri.* 1882. *No.* 5,303.

Treats with sulphate of alumina, milk of lime, and organic chlorides. The organic chlorides to be used are not named, but they are said to be derived from chloride of lime, muriatic acid and raw spirit, which appear to be added to the sewage. The inventor then filters over peat.

In England the use of "raw spirit" in sewage treatment is inadmissible on financial grounds, whatever may be the deodorising powers of the organic chlorides thus generated. It must not be forgotten that alcohol is rarely absent in sewage.

309. *Peter Jensen.* 1882. *No.* 5,536.

Uses sulphate of alumina, and then irrigates or filters.

310. *R. Nichols.* 1882. *No.* 5,981.

Lets the sewage settle; adds slaked lime, and, if extreme purity is required, sulphate of alumina.

311. *James Young.* 1883. *No.* 332.

Obtains a partial vacuum, so as to enable sewage to be boiled at a low temperature for the distillation of ammonia.

312. *James Young.* 1883. *No.* 434.

Further improvements in the separation of ammonia from sewage by distillation.

313. *W. C. Sillar and J. W. Slater.* 1883. *No.* 1,144.

Prepare crude muriate of alumina or muriates of alumina and iron by bringing solutions of sulphate of alumina (natural or artificial), and either alone or in conjunction with sulphate of iron, into contact with solutions of chloride of calcium. The clear liquid, muriate of alumina or muriate of alumina and iron, is drawn off from the deposit of sulphate of lime and used in the treatment of sewage, either alone or in conjunction with carbonaceous matter, clay, etc.

314. *J. H. Kidd and T. J. Barnard.* 1883. *No.* 1,522.
Propose mechanical arrangements for dealing with sewage.

315. *J. Bock.* 1883. *No.* 3,255.
Treats sewage with fibrous matter, salts of iron and magnesia, and effects precipitation with milk of lime. The fibrous matter employed is " wood-pulp, or paper fibre," or " fibriferous mud." There is no mention of peat.

This patent is rational in so far as precipitants and absorbents are here brought into conjoint action. But wood-pulp and paper fibre are too costly for sewage treatment. The former, containing 50 per cent. moisture, is quoted at present at £6 10s. in Hull.

316. *A. Goldthorpe.* 1883. *No.* 3,914.
Treats impure waters with phosphate of soda.

It would seem that Mr. Goldthorpe is in possession of a much more efficient method of purifying waters for industrial uses, *e.g.*, for dyeing. This process he works as a secret. The above quoted patent was never completed.

317. *John Young and Peter Fyfe.* 1883. *No.* 4,571.
Propose arrangements for separating solid and liquid matter in sewage, street slops, and smiliar material.

318. *G. Epstein.* 1883. *No.* 5,436.
Does not refer to sewage.

319. *J. H. Kidd.* 1883. *No.* 5,550.
Takes dried sewage-sludge or dried excrement, adds to it as much sulphuric acid as it will absorb, and " places it in trays in a purifier " so as to allow the gas which is generated in the retorts after cooling to be passed through such purifying materials for the purpose of fixing any free ammonia that might otherwise pass away with the gas and be lost.

The "purifier" here mentioned is a gas-purifier, and the gas generated in the retorts is coal-gas, which is passed after cooling over the sludge or excreta.

320. *T. D. Harries.* 1883. *No.* 5,147.

Mechanical arrangements for preventing the pollution of rivers.

321. *F. Herbert.* 1883. *No.* 5,850.

Proposes an electrolytic process for the treatment of sewage. Dispenses with chemicals, filter-beds, etc.

322. *S. C. Dean.* 1884. *No.* 93.

Filters sewage and foul waters through a mixture of 500 lbs. coal (not anthracite), 56 lbs. clay, 20 lbs. tar, 5 lbs. marine shells, 1 lb. borax, or 4 lbs. soda, coarsely ground together and carbonized in a gas-retort or coaking-oven. When used as filters to aerate or to force a purifying gas through foul waters, the bottom to be laid with perforated pipes, and air, steam, or a gas forced by any known means through the filter.

He also provides receiving tanks and precipitates the solid matter by a simple process, saving the ammonia and the phosphoric acid in a semi-soluble state, suitable for the food of plants, and passes the supernatant water through the filters. The "simple process" is unfortunately not given.

323. *F. Hille.* 1884. *No.* 1,279.

Intercepts floating matter by means of screens, and treats with chloride of magnesium, chloride of calcium, chloroxide of calcium or oxychloride of calcium, or perchloride of iron, or other chlorides or alum, followed by lime-water.

The application for this patent was successfully opposed.

324. *James Nuttall, F. Nuttall, and J. Rouse.* 1884. *No.* 3,417.

An arrangement to purify noxious or gaseous

vapours (!) arising from heated liquids or other materials, applicable chiefly to cooking stoves.

325. *J. Foulis.* 1884. *No.* 4,202.

Proposes certain mechanical arrangements.

326. *S. K. Page, C. E. Robinson, and W. Stevens.* 1884. *No.* 7,198.

Propose improvements in apparatus for use in the treatment of sewage, and in other operations in which solids have to be added to fluids in definite proportions.

327. *J. F. Johnstone and J. B. Alliott.* 1884. *No.* 7,387.

Arrangements for drying animal refuse and for concentrating liquids.

328. *C. Waite.* 1884. *No.* 7 481.

A process for upward and lateral filtration ; no special filtering materials proposed ; lime is applied to the deposit.

329. *W. Anderson.* 1884. *No.* 7,665.

Upward filtration through spongy iron or finely divided iron.

330. *John Hanson.* 1884. *No.* 9,587.

Claims treatment of solid and liquid impurities or foul matters with black ash waste or a liquor or decoction resulting or prepared therefrom.

This patent was successfully opposed.

331. *W. Lloyd Wise.* 1884. *No.* 9,768.

This patent does not refer to sewage.

332. *D. Nicoll.* 1884. *No.* 10,275.

Proposes certain mechanical arrangements for the treatment of sewage.

333. *J. C. Stephenson.* 1884. *No.* 10,648.

Ignites mixtures of clay and carbon and uses the product for deodorising purposes.

334. *E. J. Leveson and J. W. Slater.* 1884. *No.* 11,641.

Dissolve Kimmeridge blackstone shale or blackstone

carbon, or other shales containing alumina soluble in muriatic acid, and a considerable quantity of carbon in muriatic acid, and use the solution or mixture for the treatment of sewage along with a small proportion (1 per cent. calculated on the solids) of chloride of copper, clay and carbon, and use the mixture in the treatment of sewage.

335. *W. Astrop.* 1884. *No.* 11,901.

Precipitates sewage with "calcium sulphate and carbonate of alumina together or mixed with other ingredients." Among these ingredients there may be, it appears, alkaline sulphates, also ground cinders and ashes. There are also arrangements for drying and pulverising the deposit.

On this process it must be remarked that calcium sulphate (sulphate of lime) is not desirable, that the alkaline sulphates and ashes are at best useless, and that the existence of carbonate of alumina is not proven.

336. *G. Jones and J. C. Bromfield.* 1884. *No.* 11,971.

Treat sewage with slate-waste ground in water to an inpalpable powder.

337. *Curzon and Jones.* 1884. *No.* 12,054.

Calcine slate-waste, grind to powder and treat with sulphuric acid in heat, thus obtaining a crude sulphate of alumina, which is, of course, applicable in the treatment of sewage.

The inventors seem to forget that the cost-price of sulphate of alumina turns on the sulphuric acid. Slate-refuse is doubtless cheap and abundant, but so are many other materials containing alumina, so that there is here no distinct advantage.

338. *W. Donaldson, Isaac Shone, and E. Ault.* 1884. *No.* 12,643.

Propose to purify sewage by forcing through it atmospheric air or sulphurous acid gas or chlorine.

Similar proposals will be found in the earlier portion of this list.

339. *W. G. Gard.* 1884. *No.* 12,713.

Proposes to disinfect and precipitate sewage with a mixture of alum, sulphate of iron and sulphate of copper.

340. *J. W. Slater.* 1884. *No.* 12,830.

Treats sewage with dense and compact peat, ground up with clay and water to a thin paste, and further with slags or slag-wool treated with muriatic acid, so as to form hydrated aluminium chloride. Adds, under certain circumstances, chloride of copper or waste manganese from chlorine stills.

341. *W. D. Curzon and G. Jones.* 1884. *No.* 13,327.

Treat sewage with a double sulphate of alumina and iron obtained from "impure silicate of alumina containing iron, such as slate débris, shales, schists, and such like substances." Such minerals are roasted at a low red heat, ground, boiled in sulphuric acid, the mass lixiviated with water, the liquid run off and evaporated to a cake. This cake is mixed with $\frac{1}{4}$ its weight of slaked lime, made into a liquid of creamy consistence with water and used in the treatment of sewage.

The following questions here arise : How does the cake above described differ from Spence's "aluminoferric cake," well known in commerce ? What is the advantage of the above described process of making a double sulphate of alumina and iron over that commonly followed for making crude sulphate of alumina from shales and schists, and how indeed do they differ ? Will not the addition of slaked lime to the cake decompose *pro tanto* the sulphates of alumina and iron ?

342. *A. Angel.* 1884. *No.* 13,818.

Calcines together a mixture of clay and carbonaceous

material, and uses the product along with lime as a filtering mass.

343. *J. Y. Johnson.* 1884. *No.* 14,822.

Purifies water by saturating it under pressure and whilst in motion with compressed air, oxygen or ozonised oxygen.

344. *J. W. Slater and W. Stevens.* 1884. *No.* 15,810.

Prepare muriate of alumina for use in sewage precipitation by treating with muriatic acid minerals containing terhydrate of alumina. The muriate of alumina is used along with refuse carbon, carbonised peat, lignite and clay.

345. *J. W. Slater and W. Stevens.* 1884. *No.* 16,592.

Claim the use of fresh blood along with the ingredients named in No. 15,810.

346. *H. Wagner.* 1885. *No.* 629.

Precipitates with slaked lime, and proposes certain mechanical arrangements in which the novelty of the patent must lie.

347. *F. M. Lyte.* 1885. *No.* 900.

Uses in the treatment of sewage a soluble aluminate, preferably that of soda, followed up, if necessary, by some acid, such as sulphuric or hydrochloric, or by an acid salt. Adds further, for producing a purer water, charcoal from seaweed or peat, or shale ground together.

This is a good process. Its practical applicability turns on the price at which aluminate of soda is procurable.

348. *Y. W. Barton.* 1885. *No.* 1,345.

Proposes vehicles for conveying away town refuse.

349. 1885. *No.* 1,421.

No final specification has been filed. In this case, under the existing law, the provisional is not published.

350. *J. Richmond and T. Birtwistle.* 1885. *No.* 1,891.

Furnaces for consuming town refuse.

351. *C. Price and H. Cleave.* 1885. *No.* 2,728.
Propose mechanical arrangements for treating sewage.

352. *W. S. Page.* 1885. *No.* 2,804.
A device for lighters to convey away sewage-sludge.

353. *W. F. B. Massey-Mainwaring and J. Edmunds.* 1885. *No.* 2,885.
Force into sewage water supercharged with oxygen, and thus effect the oxidation, or, in fact, combustion of the organic impurities.

354. 1885. *No.* 2,913.
Not published.

355. 1885. *No.* 3,172.
Not published.

356. 1885. *No.* 3,682.
Not published.

357. *A. Engle.* 1885. *No.* 4,046.
Submits night-soil to destructive distillation, and utilizes the gases given off during the process.

358. 1885. *No.* 4,207.
Strains sewage through graduated screens, precipitates the clear liquid with lime, and extracts the oily and fatty matters by treatment with carbon disulphide (sulphuret of carbon).

359. *T. M. Lownds.* 1885. *No.* 4,502.
To prevent nuisance covers the liquid in urinals with a layer of oil, to which is added some disinfecting matter, such as carbolic acid.

360. *J. W. Slater.* 1885. *No.* 4,532.
Precipitates sewage with natural waters containing sulphate of alumina or other aluminous salt. If such waters are too weak, adds to them salts of iron or alumina.

Or adds sulphate of alumina to ferruginous waters. Uses along with any of these waters clay, refuse carbon, etc.

361. *J. Homes.* 1885. *No.* 5,153.

Incinerates night-soil in a special furnace (cremator), and uses the product as a precipitant, along with lime, salts of protoxide and peroxide of iron, or " acid extract of indigo, cudbear, sulphuric acid, sodium sulphate, hydrochloric acid, bichromate of potash, picric acid and logwood chips."

Bichromate of potash, as poisonous to man and beast and as harmful to vegetation, is quite inadmissible in the treatment of sewage. The advantages to be derived from the use of such bodies as extract of indigo, cudbear, picric acid, and logwood chips is exceedingly problematical, especially if we consider their prices.

362. *C. Lehoffer.* 1885. *No.* 5,227.

Proposes mechanical arrangements.

363. *M. Maehnsen.* 1885. *No.* 5,348.

Precipitates with phosphoric (basic) slags.

364. 1885. *No.* 5,448.

Not published.

365. *E. Burton.* 1885. *No.* 6,053.

Proposes furnaces for destroying excreta and refuse.

366. *F. M. Lyte.* 1885. *No.* 6,054.

Referring to former specification (A.D. 1885, No. 900), claims certain neutral salts such as calcium, chloride and sulphate, salts of magnesia, zinc, etc., instead of acids and acid salts for following up aluminate of soda.

367. *W. B. G. Bennett.* 1885. *No.* 6,055.

Mixes "carbonised clay," or "carbonised earthy deposit containing iron, and alumina," grinds to fine powder, treats with oil of vitriol, and afterwards adds water at from 150° to 212° Fahr., and uses it in sewage treatment.

368. *W. Grimshaw.* 1885. *No.* 6,453.

Dries pail-stuff, passing the fumes through current of sulphurous acid.

369. 1885. *No.* 6,559.
   Not published.
370. 1885. *No.* 6,692.
   Not published.
371. 1885. *No.* 7,186.
   Not published.
372. *B. D. Healey.* 1885. *No.* 7,703.
   Furnaces for burning up refuse.
373. *P. Smith.* 1885. *No.* 7,714.
   Arrangements for separating solids and semi-solids from sewage.
374. *J. M. H. Munro, S. H. Johnson, and C. C. Hutchinson.* 1885. *No.* 7,759.
   Treat sewage with basic cinder, followed up with milk of lime.
   Without special provision to the contrary, the effluent from this process will be alkaline.
375. 1885. *No.* 8,058.
   Not published.
376. *H. H. James.* 1885. *No.* 8,084.
   Forces air and steam into sewage.
377. *J. B. Alliott.* 1885. *No.* 8,183.
   Arrangements for dealing with mud, slops and road sweepings.
378. 1885. *No.* 8,690.
   Destructors for town refuse. Burns gases given off.
379. *J. Berger Spence.* 1885. *No.* 8,912.
   Passes sulphurous acid gas through or over the sewage in the sewers.
380. 1885. *No.* 9,268.
   Not published.
381. 1885. *No.* 9,392.
   Not published.

382. 1885. *No.* 9,992.
Not published.
383. 1885. *No.* 10,107.
Not published.
384. 1885. *No.* 10,832.
Not published.
385. 1885. *No.* 10,879.

If precipitation is not to be followed by filtration or irrigation uses per gallon of sewage a mixture of 5 to 10 grains of coke, 2 to 5 grains of clay, 2 to 5 grains of lime, and 2 to 5 grains of sulphate of alumina or of copperas. If the effluent is to be filtered or passed over land the formula used is merely 5 to 20 grains of coke and 1 to 10 grains of clay per gallon.

These mixtures will each give an alkaline effluent.

386. 1885. *No.* 11,155.
Not published.
387. 1885. *No.* 11,436.

Treats sewage with a mixture of soot—5 to 10 grains soot, 2 to 5 grains clay, and 2 to 5 grains lime, along with the deposit from the mineral spring of Southbourne-on-Sea or with a quantity of the water. Such water, or the sediment, contains a considerable proportion of sulphates of iron and alumina.

This patent was objected to on the ground of want of novelty.

388. 1885. *No.* 11,491.
389. 1885. *No.* 11,542.

Uses for sewage treatment a so-called "vitriolised ash." He forms a chloro-sulphate or sulphite of alumina and iron by heating "sulphury material," which may be ashes, spent pyrites, etc., with common salt and sulphuric acid (chamber acid) to a temperature of about 200° Fahr. This mixture is used in sewage treatment along with lime.

390. 1885. *No.* 11,750.
Not published.

391. *J. Hanson.* 1885. *No.* 12,261.
Uses vat-waste or gas lime oxidised naturally or artificially, along with lime.
This patent was objected to on the ground of want of novelty.

392. *J. Webster.* 1885. *No.* 12,344.
Treats the drainage from tank-waste, but makes no reference to sewage.

393. 1885. *No.* 12,674.
Not published.

394. 1885. *No.* 13,281.
Not published.

395. 1885. *No.* 13,587.
Not published.

396. *W. C. Sillar.* 1885. 13,749.
In treating sewage uses hot solutions of salts of alumina, etc., with or without other suitable agents.

397. *J. W. Slater and the Native Guano Company, Limited.* 1885. *No.* 13,750.
Treat sewage with a muriatic solution of copper slag in conjunction with a solution of iron slag in muriatic acid, or of hydrated aluminium chloride or manganese chloride, with or without carbonaceous matter and clay or earth.

398. 1885. 13,761.
Treats sewage alternately with an alkaline and an acid extract of Redonda, Alta-Vela or other natural phosphate of alumina. The alkaline extract is obtained by fusing the ground phosphate with caustic alkali, and the acid extract by treating the ground mineral with sulphuric or hydrochloric acid or a mixture of both.

399. *J. Carey.* 1885. *No.* 13,866.
   Barges for conveying away sewage mud.
400. *J. Carey.* 1885. *No.* 16,025.
   Ships for conveying sewage.
401. *Miller.* 1886. *No.* 10.
   Dries sewage by propelling air through it by atmospheric pressure into a partial vacuum.
402. *Botham.* 1886. *No.* 530.
   Proposes an apparatus for extracting sewage sludge out of tanks.
403. *Hardley.* 1886. *No.* 572.
404. *G. H. Leane.* 1886. *No.* 1,053.
   Filters sewage through burnt Kimmeridge shale; claims certain arrangements of plant.
405. *Cox and Cox.* 1886. *No.* 1,259.
406. *Jeyes.* 1886. *No.* 1,261.
407. *Hartland.* 1886. *No.* 1,727.
408. *F. Candy.* 1886. *No.* 1,792.
   Ignites in closed vessels a mixture of quicklime and tar or tarry oils, resins and bitumens or other liquid or semi-liquid carbonaceous matters, and uses the product in the treatment of sewage. This mixture is an attempt to combine the purifying agency of carbon with lime as a precipitant.
409. *Davis.* 1886. *No.* 1,840.
410. *F. H. Danchell.* 1886. *No.* 2,439.
   Chars sewage mud or deposits, which he then calls charcoaline, and filters sewage through it.
411. *Cobley.* 1886. *No.* 2,752.
412. *J. B. Hannay.* 1886. *No.* 3,217.
   Precipitates sewage with a mixture of lime or carbonate of lime and clay and makes cement of the deposit.
   Here, of course, as in the Scott processes the plant-food present in the sewage is wasted.

413. *J. W. Slater, S. K. Page, W. Stevens, and the Native Guano Company, Limited.* 1886. *No.* 3,973.

Prepare a sulphate of manganese by gently igniting in a crucible a layer of sulphur over which is placed a layer of the black oxide or peroxide of manganese. Or they prepare mixed sulphates of manganese and alumina by treating in a similar manner a mixture of peroxide of manganese and Eglington clay, bauxite or other hydrates of alumina. Or they prepare chloride of manganese or of manganese and aluminium by adding to the aforesaid mixtures common salt. The products are used in the treatment of sewage alone or in combination.

414. *L. G. G. Daudinart.* 1886. *No.* 4,203.

Treats sewage first with lime and then with a mixture of sulphate of alumina and chloride of zinc.

The well-known poisonous character of the salts of zinc renders them very undesirable in the treatment of sewage. If the zinc is precipitated the sludge will be unsafe as manure for plants, and if any of it remains in solution the effluent cannot be fit for admission into streams.

415. *F. Petri.* 1886. *No.* 4,512.

Precipitates first with lime, and then with epsoms or other salts of magnesium or barium, and filters the effluent finally through matters containing tannin.

Great care will be here necessary to prevent barium remaining in solution. Nor will its presence in the sludge be advantageous for manurial purposes, since it is not a plant food. Matter containing tannin, if used as a filter, is well suited for removing certain impurities from the effluent water.

416. *T. Reid.* 1886. *No.* 4,544.

Mechanical arrangements for filtering.

417. *W. F. B. Massey-Mainwaring.* 1886. *No.* 4,878.
Proposes improvements in filter presses for drying sewage sludge, &c.

418. *G. R. Redgrave.* 1886. *No.* 6,520.
A lime process for the conversion of sewage matters into cement.

419. *T. H. Cobley.* 1886. *No.* 6,732.
Uses the ash of sewage deposits in place of lime as an addition to sludge for pressing.

Long experience has shown that there is not the least occasion for an addition of lime to sewage mud before pressing.

420. *J. Fenton.* 1886. *No.* 7,333.
Passes sewage through a tank containing alum, and then into strainers.

421. *Alsing.* 1886. *No.* 7,730.

422. *H. R. Newton.* 1886. *No.* 8,144.
Proposes certain mechanical arrangements.

423. *Georgi.* 1886. *No.* 8,411.

424. *F. H. Danchell.* 1886. *No.* 8,469.

425. *James Bannehr.* 1886. *No.* 8,874.
Claims certain mechanical arrangements.

426. *Bohlig and Hayne.* 1886. *No.* 9,276.
Produce "carbon magnesia" from chlorate of magnesia. Chlorates as waste products are not very abundant.

427. *H. W. Lafferty.* 1886. *No.* 9,431.
Propose to utilise the refuse of breweries and distilleries.

428. *W. Burns.* 1886. *No.* 9,569.
Mixes animal dung (is there any other than animal dung?) with sawdust, peat, coal dust, chalk, blood and iron to form a "depurating carbon," and distils the mixture in retorts.

429. *H. Cobley.* 1886. *No.* 9,847.

430. *W. Astrop.* 1886. *No.* 10,047.
Proposes drying arrangements for sewage sludge.

431. *Johnstone.* 1886. *No.* 11,068.

432. *Hannay.* 1886. *No.* 11,165.

433. *W. P. Thompson.* 1886. *No.* 11,366.
Proposes apparatus for drying "offal." There is no direct reference to sewage.

434. *A. Forrest and W. Welsh.* 1886. *No.* 11,409.
Propose apparatus for drying up blood, sewage, &c., so as to be available for manure.

435. *Miller.* 1886. *No.* 11,461.

436. *G. H. Leane.* 1886. *No.* 11,820.
Uses Kimmeridge carbon in place of lime, as an addition to sewage mud for pressing.

Kimmeridge carbon is undoubtedly much preferable to lime, but no addition at all is necessary.

437. *Gerson.* 1886. *No.* 11,830.

438. *J. H. Barry.* 1886. *No.* 11,833.
Strains off the liquid portion of sewage from the deposits, concentrates the liquid by distillation, returns the residue to the deposit, and treats it by precipitation and filtration.

No novel precipitant is claimed. Without doubt a very good effluent can be obtained in this manner, provided a good precipitant is selected. The ammonia given off during the distillation may be collected, and either added to the precipitate, or may be otherwise utilised at pleasure. But the process must prove expensive, as the straining and the distillation are superadded to the ordinary operations of precipitation treatment.

439. *Soldenhoff.* 1886. *No.* 12,259.

440. *Hallett.* 1886. *No.* 12,382.

441. *Bremner.* 1886. *No.* 12,981.

442. *Perkins.* 1886. *No.* 12,850.

443. *Butterfield and Mason.* 1886. *No.* 13,007.
444. *Abel.* 1886. *No.* 13,470.
445. *Ames.* 1886. *No.* 13,791.
446. *F. Candy.* 1886. *No.* 13,829.

Prepares a compound which he names " magnetic precipitant," and which he uses in the treatment of sewage. He ignites in a closed vessel argillaceous iron stone or iron ores (which should contain either lime or magnesia, or both of them) and either with or without the addition of carbonaceous matter. The temperature must not be high enough to vitrify the matter, which is afterwards treated with sulphuric and hydrochloric acid.

The final product is, of course, a mixture of carbon with sulphates of alumina, iron, lime, and magnesia, or with the corresponding chlorides. No one can deny that the combination of salts of iron and alumina with carbon form an efficient precipitating mixture. But whether there is any advantage in preparing such mixture in the manner specified, is an open question.

447. *Donnithorne.* 1886. *No.* 14,221.
448. *Burns.* 1886. *No.* 15,222.
449. *Mayes.* 1886. *No.* 15,341.
450. *W. F. Mast.* 1886. *No.* 15,887.

Adds to sewage matters an alkaline base, such as lime and common salt, or some equivalent chloride. He then heats to high temperatures in a suitable vessel, and collects the ammonia.

The novelty and the value of this process turn on the alleged effects of the addition of an alkaline chloride, such as common salt, in expelling ammonia from organic matter.

451. *Hartland.* 1886. *No.* 16,039.
452. *Wilson.* 1886. *No.* 16,244.
453. *Brown.* 1886. *No.* 16,461.
454. *Barry.* 1886. *No.* 16,866.

CHAPTER XX.

THE DISCUSSION ON DR. TIDY'S PAPER, HELD MAY 5TH, 1886, BEFORE THE SOCIETY OF ARTS.

I AM prevailed upon to notice this discussion, not because it brought to light anything at once novel and true—consisting, indeed, as it did, in no small part, of matter possessing neither of these attributes—but because it affords good specimens of the views still current in official circles. I regret having to put on record the fact that two, perhaps more, of the speakers who took part in the debate were apparently reading prepared and, to some extent, independent papers on sewage treatment, rather than simply criticising Dr. Tidy. This is a defect not readily avoidable in the case of an adjourned discussion.

It will be remembered that Dr. Tidy did not, in his original paper, pose as the advocate of *any* one exclusive system. But this lack of an Athanasian Creed has left him open to criticism from all sides. The exceptions taken to his views, and his arguments in the discussion in question, came mainly from irrigationists and filtrationists. But room is left for the friends of precipitation, as rationally conducted with the aid of absorbents, to say a word.

The learned doctor was not seen to draw a sufficiently broad distinction between precipitation pure and simple, *i.e.*, the treatment of sewage by means of salts of alumina, iron, magnesia, etc., separately or jointly, with or without the addition of lime ; and, on the other hand, precipitation *plus* absorption, the absorbent materials being added first, and the true precipitants afterwards. Those who advocate and practice this double method have the advantage over the irrigationist, inasmuch as, like him, they absorb the soluble matter of the sewage in earth or clay, etc., but are able constantly to bring fresh material into play, so that no choking or supersaturation can ever arise. They have the advantage over the precipitationist " pure and simple," inasmuch as all the doubts and queries and quibbles about " clarification, but not purification," pass by them like very idle wind. With the addition of the absorbent materials to the sewage the bad smell disappears, and on the subsequent introduction of the precipitant the suspended sewage matters, as well as the particles of the absorbent, now saturated with the soluble and gaseous impurities, are thrown down together. In short, we have first purification and then clarification.

When Dr. Tidy's paper was read there was to be seen a press-cake of sewage deposit. Any person acquainted with sewage treatment would have known at once by the smell that this cake was the result of some modification of the lime process. But, so far as I can remember, the lecturer said no word to that effect; no more did any of his critics during the subsequent discussion. Hence, it was left open for any person so disposed to insinuate, or actually to believe, that this cake was a product (say) of the " A B C " process, or of some other of those processes combining absorption and precipitation. That the full truth should come to light in this matter is the more urgent in view of the somewhat strong language used in the discussion by Dr. Dupré, F.R.S.

This gentleman is reported to have said: " It was a well-known trick of counsel* in sewage cases to have a large cake

* That counsel have certainly their "tricks" in sewage cases, as well as in others, is perfectly true. But—

"For ways that are dark, and for tricks that are vain,
　　The Royal Commissioner is peculiar,
　　Which the same I am free to maintain."

of *this pressed sludge* brought into court to point out how inodorous it was, and to almost advise persons to take it home and place it on the mantelpiece ; but it was not said that that was generally especially pressed cake, very carefully dried by exposure to the air. If you took *that cake* and moistened it and bottled it up, and then smelt it again a week after, no one would think of recommending it to be kept in a dining-room. They had all heard Dr. Tidy's story about walking about amongst *thousands of tons of this which really smelt rather nice ;** but he (Dr. Dupré) had had the misfortune also to go about *this heap of thousands of tons*, having visited *it* on a cold May morning, when there had been a frost during the night—a very unfavourable time for producing smells. He visited the heap with the engineer, who told them that he could not smell anything ; but, after going some distance in the direction of the wind, they found it really very offensive. The engineer said he never smelt it so strong before, although it was a cold morning ; they went as far as they could, and measured the distance, and found it was six hundred yards, and there the smell was most powerful and offensive."

Now, I do not for one moment believe—and I should gravely deplore having to believe—that Dr. Dupré in the above remarks was knowingly and intentionally misleading his hearers. But it is, I think, to be regretted that he did not say plainly by what process the " thousands of tons " had been obtained. We must remember that :—

1. There is cake and cake.
2. That the sample on view on the occasion of Dr. Tidy's paper was a lime-process cake, and that its smell was decidedly offensive.
3. That the cake from lime processes is well known to be, in quantity, a manifest nuisance, whilst the same charge

* Dr. Tidy said, "and really smelt very little"—a very different proposition.

cannot be brought against the cake or mud produced by a combined absorption-precipitation process, or even by a simple precipitation process, where lime plays no part except that of removing any excess of acid.

Dr. Dupré might, I fear, when speaking of *this pressed sludge*, *that cake*, etc., be understood to mean any and every kind of sewage-sludge or cake. Hence I am compelled to mention, lest his " story " of going about this *heap of thousands of tons* should be misconstrued, that Dr. Dupré has never visited the Aylesbury Sewage Works. This I have ascertained by special and very careful inquiry. Had he so done, he would have found neither a " most powerful and offensive smell," nor a heap of thousands of tons of sludge or cake. Nor, so far as I can learn, is there any other place in the kingdom where a true absorption-precipitation process is at work where a nuisance is experienced, or where "thousands of tons" of sludge or cake could be met with. Hence it follows that when Dr. Dupré went a-Maying it must have been to some lime-process works, and the defenders of such processes may refute his charges—if they can.

Taking in order of time the speakers who bore part in the discussion, I am glad to note the fact that Lieut.-Col. A. S. Jones, V.C., evinced in his remarks more candour and breadth than is usually to be found among irrigationists. He seems to me to raise a question interesting and legitimate enough in itself, but which has already been practically decided. He thinks that in judging the effect produced by any given precipitant, we should compare the effluent, not with the raw sewage, but with the liquid obtained by simple subsidence without the aid of any chemicals. But the amount of spontaneous subsidence from sewages of identical composition is by no means a constant quantity. It varies with temperature, atmospheric pressure, and time allowed. Thus the liquid obtained from such natural subsidence would furnish no term of comparison with the effluent obtained by any rational precipitation process.

Further, the deposit obtained by subsidence is different, both in quantity and quality, from that produced by precipitation. The spontaneous deposit is decidedly and invariably smaller in weight, and contains a much lower proportion of organic matter. It consists principally of silt, sand, road dust, and other heavy and insoluble mineral bodies. I have repeatedly allowed quantites of sewage to subside for different lengths of time, ranging from six to twenty-four hours. When all subsidence was at an end, I have carefully decanted off the clear liquor, and on adding to it a suitable precipitant—say hydrated aluminium chloride or aluminium acetate—I have obtained a very copious precipitate. Hence the dictum that a precipitation process is little better or more efficient than a subsidence process can have sprung only from inobservance or from prejudice.

It is highly probable that if we allow sewage to settle spontaneously, and then add to the clear liquid some precipitating agent, the two deposits added together would not be equal in amount to that obtained by at once adding the precipitant to the sewage. The reasons for taking this view are:

1. Recent sewage is more easily precipitated than such as is stale.

2. During spontaneous subsidence, which is a much slower process than precipitation, fermentation sets in. Solids are converted into liquids, and both solids and liquids into gases. Simple subsidence has been proposed by not a few inventors as one step in systems of dealing with sewage, followed up in most cases by treatment with chemicals, by irrigation or by some kind of filtration. But the patentees appear to have been fully aware of the character of the fumes given off. They recommend that the tanks should be covered in, and connected with a tall chimney or with a furnace to convey away or to destroy such fumes.

In a further paragraph of his speech (or may we say paper?), Col. Jones goes, I submit, beyond the legitimate boundaries of a discussion on Dr. Tidy's paper. The learned doctor had certainly not expressed any opinion on—much less defended—the policy of the Native Guano Company. To have done so would have been on his part a departure from his subject, and a breach of method. But Colonel Jones, strangely enough, intercalates an examination of the policy of the Native Guano Company in his critique of Dr. Tidy's paper!

Of the "two serious mistakes" which Col. Jones considers that the Native Guano Company has committed, the second, as he admitted almost in the same breath, was scarcely under its control, viz., that "the Stock Exchange afforded opportunities for gambling in 'Natives.'" The shares of *all* limited companies are exposed to "bull and bear" manipulations, which neither the Legislature, nor the Courts, nor Boards of Directors can hinder or control.

Col. Jones's remarks on the scheme adopted by the Metropolitan Board of Works are judicious. He considers the effluent from such an imperfect precipitation-process "inadmissible into the estuary even at Thames Haven." Besides showing the inefficiency of the prescribed dose, he further, with perfect correctness, reminded his hearers that the deodorisation produced by the application of permanganate is transitory in effect. "This," he truly said, "anybody might ascertain by keeping a bottle of effluent deodorised by permanganate for a few days, and as the sewage is not supposed to reach the Nore for some six weeks after its discharge at Barking, it would seem hardly worth the expense of deodorising it for about one-twentieth part of that period."

Certainly, if permanganate is to be used to advantage, as much as possible of the organic matter ought to be previously removed from the sewage. Further, if the permanganate is to oxidise, in other words burn up the residual organic matter, it must be used in much greater quantity than the Metropolitan Board have proposed. Permanganate entirely fails in removing the microbia which are now so much dreaded.

Neither Col. Jones nor any following speaker succeeded in showing how sewage irrigation could be other than an evil when the amount of rainfall is in itself more than is good for the land.

Mr. Baldwin Latham considered that sewage farms were rather favourable than otherwise to public health. This statement differs greatly from the results obtained in India, as recorded by Markham. There, irrigation with ordinary river water, and applied only in dry weather, is recorded as having had an unfavourable effect on public health.

Mr. Peregrine Birch endeavoured, very unsuccessfully, to explain and justify the statement of Professor Frankland, that the sewage of midden towns was practically as foul as the sewage of closeted towns. The additional water thrown into the sewers of the latter does not remove or destroy the excrementitious matters. In a midden town all the solid excreta, and, to a very large extent, the urine also, are kept out of the sewers. That the quantity of water introduced can compensate for this pollution is a delusion.

Mr. Birch thinks that the cracks in clay lands can be got rid of by putting on more sewage, irrespective, of course, of its action on the crops. But what of the cracks and crevices in subjacent chalk and other strata? These, it is to be feared, no excess of sewage will fill up. From the conflicting statements of Way, that a light soil is best for irrigation, and of Liebig that a stiff soil is best, Mr. Birch draws the singular inference that "there was scarcely any soil in the world that would not do for the purpose." If nothing more was known on the subject than the dicta just quoted, it would be quite as legitimate to infer that " scarcely any soil in the world would do for the purpose."

Mr. Birch agrees with Dr. Tidy, and differs from Col. Jones, in thinking that "sewage, spread over the land, forms *papier mache*."

But his grand *coup*, delivered over his own knuckles, was an attempt to show "the absurdity of the assertion that native guano is worth £3 10s. a ton." As Dr. Tidy had not made any such assertion, nor discussed the value of native guano at all in his paper, Mr. Peregrine Birch was clearly out of order in introducing the subject. But let us examine if what he said could in any manner justify his departure from the question before the society. He took up a table, showing the composition and value (?) of the weekly out-put at Aylesbury, giving the following figures for sixteen tons :—

	£	s.	d.
Charcoal—4 tons 14 cwt. at . .	0	1	0
Clay—6 tons 10 cwt. at	0	1	0
Alum—1 ton 17 cwt. at	2	0	0

and consequently he finds, by way of difference, that there are in sixteen tons of native guano only two tons eighteen cwt. of sewage matter !

Here he evidently forgets or ignores the fact that the charcoal used contains at least 50, often 60, per cent. of water, which passes out in the effluent. The clay, in like manner, contains 40 per cent. of moisture, whilst of the alum only the alumina, 13 to 14 per cent. enters into the precipitate. So that instead of the 13 tons 1 cwt. of non-manurial matter which Mr. Peregrine Birch estimates as present in native guano, there are only about 6½ tons, and the proportion of matter derived from the sewage, in place of 2 tons 18 cwt., is about 9 tons 9 cwt. in a dry sample, or in one containing 14 per cent. of moisture 8 tons 2 cwt. !

But we must follow Mr. Peregrine Birch two steps further. He seems to forget that most common clays contain a considerable proportion of coarse sand, gravel, and pebbles, which remains in the grinding pans, mixing pits and shoots conveying the A B C mixture, and never reaches the precipitating tanks at all ! Hence a further deduction must be made from the estimate of non-manurial matter, and a corresponding increase of the sewage matters.

It may further be asked, what conceivable bearing the cost of the alum (alum-cake), here given as £2 per ton, has upon the value of the native guano? Alum has no manurial value at all, and the alumina which it contains acts merely by forming, with the nitrogenous matters of the sewage, a lake-like compound. But here we may very well afford to take leave of Mr. Peregrine Birch.

Next followed Dr. Percy Frankland, who considered that the strength of the sewage in midden towns, as compared with water-closet towns, is simply a question of water supply. I have already shown that fæcal matter, even *plus* a somewhat large supply of water, must give a sewage fouler than one in which such matter is absent. Says Dr. Frankland :—" The average for the water-closet towns was largely obtained from London itself, and the midden towns were principally taken from Lancashire, where the water supply per head was much smaller than in London." But, on the other hand, it should be remembered that the average rainfall of Lancashire is notably greater than that of London. The comparison is further complicated by the fact that in Lancashire it is difficult to find a town where the sewage does not receive much industrial waste waters. If the late Rivers Pollution Commissioners had wished to make a really useful comparison they should have selected two groups of purely residential towns, the one closeted, and the other provided with middens. If the contention is true, that the pollution of the sewers is simply "a question of water supply," it is hard to see the value or the relevance of the fact.

The next speaker was Dr. Dupré, a chemist of unquestionable eminence and authority. We find him venturing on the conclusion that " no precipitation process which had been at present brought forward did sensibly more than clarify the sewage." Strange to say, he added, that "in this opinion he was borne out by experiments which had been made by Dr. Tidy and Professor Dewar." Now, as Professor Dewar subsequently pointed out, "three-fourths of the readily oxidisable matter was removed" from sewage by the process in use at Aylesbury, including dissolved as well as suspended impurities ! !

Dr. Dupré further falls into error, when speaking of the A B C process, by ignoring, like Mr. Peregrine Birch, the moisture present in the materials used. Hence his statements that the pressed cake is chiefly formed of the materials which had been added, and that no real benefit was produced require to be taken with a very large grain of salt. Whence come the three per cent. of ammonia and the phosphoric acid, equal to five per cent. tricalcic phosphate, which recent analysis shows in the native guano ?

Mr. Dibdin spoke of the "common mistake" that because a glass of effluent water was clear, therefore it was pure. Clear waters, not merely effluents, are certainly not necessarily pure—a point of which the public have been by this time informed to satiety. But if the suspended matter is, as Colonel Jones remarks, "the foulest of town sewage;" if it is, as Lord Bramwell's Commission called it, the "crux" of the question, even clarification is not to be despised. There is a "common mistake" about "pure water" and "purification." Pure water, in the strictest sense of the word— and no other is legitimate—is almost as much an abstraction as the lines and points of the mathematician. It has never been obtained on a practical scale. If, therefore, we are told that "precipitation does not purify sewage," the remark is a mere truism. Nor does any one contend that sewage can be rendered fit or safe to drink, except by natural processes. But the assertion that sewage cannot be rendered by any precipitation process, safe and fit to be admitted into "any river," must be pronounced exceedingly rash.

There is very little novelty in the information that an excess of precipitants, beyond a certain limit, does not necessarily improve an effluent. I could find working men who have been aware of this truth a dozen years ago. Especially does this rule hold good if lime is concerned. Mr. Dibdin says, truly enough, that "this large quantity of lime (700 grains per gallon) dissolved some of the suspended matters, and brought them into solution. This is, in fact, one of the grave objections to the use of lime as a precipitant.

We find here a very curious report on a sample of effluent which had been examined by Dr. Bell, of the Excise Laboratory, Somerset House. It is said that "after careful examination more organic matter was found in the effluent than there was in the sewage." Whence had it come, if only lime had been added?

This critique had best close with the words of Dr. Dupré:—" Let everybody fairly co-operate in the question, and, above all, let every one give to those opposed to him credit for being actuated by the same honourable motives as he was himself." Very rarely indeed have the advocates of chemical processes received hitherto such credit, especially from official bodies.

INDEX.

A B C Process, 85, 97, 258
Absorbents in sewage treatment, 103
Aeration necessary for fish, 148
—— of effluents, 112, 121
Aire River, 11, 160
Alkaline effluents injurious, 89
Alum not good for sewage treatment, 97
Alumina acetate too costly, 88
—— Carbonate, 100
—— Muriate, 99
Aluminate of soda, 99
Aluminium, basic sulphates of, 98
—— Hydrated chloride of, 98
—— Salts of, 97
Alumino-ferric cake, 98
Ammonia in waters, 20
—— Volatilised by chalky soils, 175
Ammonium, salts of, 88
Amphibia in streams, 149
Antimony in manufacturing sewage, 4
Arable soil an absorbent, 103
Archiv fuer Hygiene, 19
—— *der Pharmacie*, 157
Arsenic in manufacturing sewage, 4
Asbestos a filtering medium, 3
Auerbach, Dr., 148
Aylesbury, sewage-manure of, 182
—— Sewage Works, 57, 85, 86, 108, 146, 260

a rium, salts of, unfit for sewage treatment, 93
Barking Creek, sewer outfall at, 36

Bazalgette system, 38, 185
Beggiato à alba, 144, 151
Bell, Dr., 267
Birch, Mr. Peregrine, 263
Blackburn, lime process at, 91
Bleaching liquors kill fish, 19, 93
Boehmer, Dr., action of lime on sewage, 90
Bottles for sampling, 170
Brautlecht, Dr., precipitates germs with alum, 99

Cake alum, 98
Cake, sewage, 258
Calder and Hebble Navigation, 138
Carbolic acid unsafe in waters, 76
Carbonates unfit for sewage treatment, 101
Carbons, waste, 105
Cement, cf doubtful value, 105
Cement processes, Gen. Scott's, 125
Cesspools, defects of, 25
Charcoals, 104
Chemical News, 175
Chemiker Zeitung, 157
Chiltern Hills water, 146
Chinese method of purifying water, 99
—— Adopted by French in Tonkin, 99
Cholera, germs of, distributed by flies, 62
Clay, fatty, an absorbent, 103
Climate, its bearing on irrigation, 47
Coal-ashes of little use, 104
Coal-tar proposed, 94
Coke said to remove microbia, 104

INDEX. 269

Commission, Royal, on Metropolitan Sewage, 86
—— Refuses evidence, 85
—— On Rivers Pollution, 86, 93, 160, 161
Commissioners on Rivers Pollution, their Recommendations, 159
Confervæ improve effluents, 166
Copper, use of, 102
Coventry, sewage treatment at, 70
Cresswell, Mr. C., 145
Crooskes, Mr. W., F.R.S., 115,164
Crosness, sewage outfall at, 41
Cylinders for drying sludge, 177

Danzig, well waters of, 157
Denton, Mr. Bailey, 79
Deodorising sewage, 119
Dewar, Professor, 85, 266
Dibdin, Mr., 266
Disinfection applied to sewage, 70
Dupré, Dr., F.R.S., 258, 264, 265, 266
Dust-bins, evils of, 37
Drying cylinders, 177

Effluents, alkaline, bad, 89
—— Examination of, 164, 166
—— Improved by green vegetation, 166
Elodea Canadensis, 147

Filters, cleaning, 75
—— Construction of, 73
—— Duties of, 72
—— Materials for, 74
Filtration, how differing from irrigation, 72
—— Intermittent, 78
—— Upwards and downwards, 72
Fish, substances hurtful to, 19
Flies, carriers of infection, 37, 60
Frankland, Professor E., 53, 79, 80, 90
—— Dr. Percy, 55, 92, 104
Froth in tanks and channels, 167

Gasworks, refuse may not be put in sewers, 87

Gennevilliers, bad smell at, 57
—— Irrigation farm at, 49
—— Sewage at, 154
Gerardin, M., 137
Germs not removed by irrigation or filtration, 55
—— Thrown up from still waters, 59
Gesundheits Ingenieur, 140
Gibbs, Mr. W. A., on sewer gas, 36
—— Harvest saver, 49
—— and Borwick, drying cylinder, 177
Gnats, larvæ of, in water, 150
Grassi, on flies as carriers of disease, 62, 63
Ground water, 34
Gypsum useless in sewage treatment, 93

Hendon, use of lime recommended at, 91
Hertford, sewage treatment at, 203
Hogg, Mr. Jabez, M.R.C.S., F.R.M.S., 56, 77
Hulwa, Dr. F., 140
Hypochlorites not fit for use in treatment, 93

India, irrigation in, 58, 263
Intermittent treatment, 107
Irk, River, 11
Irrigation efficacious, 45
—— Inverse, 107
—— Modifications of, 68
—— Quality of soil for, 46
—— Retards crops, 50
—— Made self-supporting, 49
—— Suitable climate for, 47
Irwell, River, 11

Jackson, Mr. L. d'Aguilar, C.E., 16
Jones, Colonel, V.C., 260

Kelvin Water, 160
Kimmeridge carbon and blackstone, 74
Kingston, fish at sewer-mouth 149
Knostrop Sewage Works, 110
Koch, Dr., 156

INDEX.

König, Prof., action of lime on sewage, 90

Land waters polluted, action with sea water, 43
Landwirthschaftliche Jahrbücher, 90
Latham, Mr. Baldwin, 263
Lead unfit for treating sewage, 88
Leeds, Dr. A. R., 139
Leeds, sewage of, 154
Lefeldt, Herr, opinion on irrigation, 48
—— On smells at Romford, 57
—— On unassimilated sewage in grass, 65
Leicester, lime process at, 91
Liebig, on soils fit for irrigation, 263
Liernur, Captain, 29
Lignite, absorbent and filter material, 113
Lime, its action on dyes in sewage, 90
—— Effluents and muds, 90
—— Hurtful to fish, 20, 90
Lime process at Blackburn and Leicester, Prof. Frankland, on, 91
London, sewage of, 154
Lyte, Mr. F. Maxwell, 99

Macadamized roads deteriorate sewage, 3
McCarter, Judge, on nuisances, 9
Maddox, Dr., flies conveying comma bacillus, 62, 63
Maercker, Dr., on nitrogen in waste waters, 53
Manganese a good precipitant, 101
Manson, Dr., on mosquitoes as carriers of germs, 63
—— Mr. E., C.E., 54
Markham, Mr., on irrigation in India, 58, 263
Marsh-gas, in sewers, 17
Melbourne, sanitary state of, 32
Microbia, 10, 11
Micro-organisms, 10, 11
"Milk in Health and Disease," 65
Moor-earth, as filtering medium, 76

Mud from sewage, drying, 173
Mussels rendered poisonous by sewage, 41

Native Guano Company, 262
Nesbit, Mr. A. Anthony, on bleach liquors in rivers, 93
Night-soil applied to fields, 63
Nuisance, definition of, 9, 57
Nuneaton, sewage treatment at, 202

Oder, River, 140
Odling, Professor, 164
Officials at sewage works, hints to, 168
Outfall channels, 167
Oxygen, free, generally absent in sewage, 17
Ozone as purifying agent, 219

Page, Mr. S. K., 76
Paris, sewage of, 154
Passaic River, 139
Pasteur and Chamberland, their filter, 76
Patents, " bogus," 116
Peat as absorbent, 104
Peat Engineering Company, 76
Percy, Earl, his Sewage Bill of 1885, 158, 189
Petroleum proposed for treating sewage, 94
Pettenkofer, Prof. von, 34
Phosphate processes, 95
Phosphate Sewage Company, 203
Pollution, subterranean, 158
Purification, indicated by blue tint 115
Precipitants, use of hot, 97
—— Used in excess, 167
Precipitation *plus* absorption, 257
—— Control of, 114
—— Re-treatment in, 115
Press-cake, limed, 258
Ptomaines, 41, 94

"Recommendations" of Rivers Pollution Commission, 159
—— Criticised, 160
Rivers Clarification Company of Leeds, 221

INDEX.

River Purification Association, 203
Robinson, Prof. H. (of Dublin), 77
Rye-grass converted into hay, 49

Saare and Schwah, observations on substances deadly to fish, 19
Salt, common, in waters, 15
Sampling effluents, precautions in, 170
Sanitary Institute, 77
Sanitary Record, 77
Schutzenberger and De Lalande, 137
Scum, oily on waters, 158
Seine, River, 138
Settling pits, 68
Sewage, discharge of into sea, 40
——— Filtration of, difficult, 3
——— Flow of, in dry weather, 7
——— Fungus, 151, 153
——— Of manufacturing towns, 2
——— Nature and composition of, 1
——— Of Residential towns, 3
——— Alkaline, 18
——— Specific gravity of, 115
——— Where injurious, 9
Sewerage, single and double, 7
Sewer-gas, 33
Sewers, ventilation of, 18
Silica, gelatinous, absorbent, 105
Sillar, Mr. W. C., 97
Silt and road-dirt, 3
Slag, basic, added to mud, 175
Smee, jun., Mr., his experiments, 65
Smell, offensive, not always found in putrid matter, 16
Smell, sense of, 165
Smorbo, 196
Soap in sewage, 51
Soil, its sanitary efficacy, 28
Soot of little value, 105
Spence, P. (the late), his alumino-ferric cake, 98
——— his central chimney for sewer-gas, 36
Spongy iron as filter material, 74

Stevenson, Dr., 91
Storm-water, 110
Streams, pure, insects in, 150
Strontium, salts of, admissible, 94
Sulphur proposed for use, 94
Sulphurets rarely admissible, 93

Tanks, construction of, 111
——— Covered, a mistake, 111
——— Foul, 112
——— Room, ample, 110
Thames Haven, proposed out-fall, 42
Tidy, Dr., 85, 257
Tonkin, French troops in, 99
Towns, closeted, sewage of, 6
——— Residential, sewage of, 2
Turpentine proposed for use, 94

Urea, its presence transitory, 14

Vesle, River, 137
Visitors at sewage works, 169

Wanklyn, Prof. J. A., 127
Water, consumption of by closets, 32
——— Pure, does not occur in nature, 85
——— Purification, Chinese method, 99
——— Samples of, fraudulent, 169
——— Softening, Clark process, 81
——— Sparkling, 158
——— Supply, effect on sewage, 7
Waters, chemical analysis of, 157
——— Impure, why sometimes harmless, 10
——— Dangerous externally, 16
——— Microscopic examination of 157
Way, T., on soils fit for irrigation, 263
Well-waters, 157

Zinc, salts of condemned, 102

www.ingramcontent.com/pod-product-compliance
Lightning Source LLC
Chambersburg PA
CBHW032107230426
43672CB00009B/1657